Oscillation and Waves: Fundamental Concepts in Physics

Oscillation and Waves: Fundamental Concepts in Physics

Editor: Paula Willoughby

New York

Published by NY Research Press
118-35 Queens Blvd., Suite 400,
Forest Hills, NY 11375, USA
www.nyresearchpress.com

Oscillation and Waves: Fundamental Concepts in Physics
Edited by Paula Willoughby

International Standard Book Number: 978-1-63238-606-9 (Hardback)

Cataloging-in-Publication Data

Oscillation and waves : fundamental concepts in physics / edited by Paula Willoughby.
p. cm.
Includes bibliographical references and index.
ISBN 978-1-63238-606-9
1. Oscillations. 2. Waves. 3. Physics. I. Willoughby, Paula.
QC151 .083 2018
531.32--dc23

Contents

Preface

The repetitive variation in time of a motion from one place to another or towards equilibrium point is known as an oscillation. Mechanical oscillations are called vibrations and other forms of oscillations are pendulum and alternating current power. The different types of oscillators are harmonic oscillators and anti-vibration compound. When oscillations travel via mass or space carrying or transferring energy they are known as waves. This book unfolds the innovative aspects of oscillation and waves, which will be crucial for the holistic understanding of the subject matter. The topics included in it are of utmost significance and bound to provide incredible insights to readers. The textbook aims to serve as a resource guide for students and experts alike and contribute to the growth of the discipline.

A foreword of all chapters of the book is provided below:

Chapter 1 - Oscillation is the repetition of an equilibrium point or various states of values, in time. It can be noticed in the beating of the heart, geothermal geysers, vibrating strings of a musical instrument, etc. This is an introductory chapter which will introduce briefly all the significant aspects of oscillations; **Chapter 2 -** The concept by which a system oscillates with greater amplitude and at specific frequencies with the application of a vibrating system or an external force is known as resonance. Some of the examples of resonance are swings, Tacoma narrows bridge, modern watches, etc. The topics discussed in this chapter are coupled oscillators, wave, sine wave and electrical circuits. The diverse applications of resonance in the current scenario have been thoroughly discussed in this chapter; **Chapter 3 -** Electromagnetic radiations are self-propagating waves of electromagnetic field containing electromagnetic energy. It is possible for electromagnetic waves to oscillate with the help of an electric dipole. This concept is very important to manufacture devices like the laser. The chapter strategically encompasses and incorporates the major components and key concepts of electromagnetic radiation and electromagnetic waves, providing a complete understanding; **Chapter 4 -** Waves can have positive and negative values. When two waves are superposed, they can generate a wave of higher, lower or same amplitude. This phenomenon is known as interference. The aspects elucidated in this section are of vital importance, and provide a better understanding of interference; **Chapter 5 -** The phenomenon where light bends due to simultaneous interference of many waves is known as diffraction. Diffraction occurs in all types of waves. It is very important in the creation of the X-ray. The topics discussed in the chapter are of great importance to broaden the existing knowledge on diffraction.

At the end, I would like to thank all the people associated with this book devoting their precious time and providing their valuable contributions to this book. I would also like to express my gratitude to my fellow colleagues who encouraged me throughout the process.

Editor

Brief Introduction to Oscillations

Oscillation is the repetition of an equilibrium point or various states of values, in time. It can be noticed in the beating of the heart, geothermal geysers, vibrating strings of a musical instrument, etc. This is an introductory chapter which will introduce briefly all the significant aspects of oscillations.

Oscillation

Oscillation is the repetitive variation, typically in time, of some measure about a central value (often a point of equilibrium) or between two or more different states. The term *vibration* is precisely used to describe mechanical oscillation. Familiar examples of oscillation include a swinging pendulum and alternating current power.

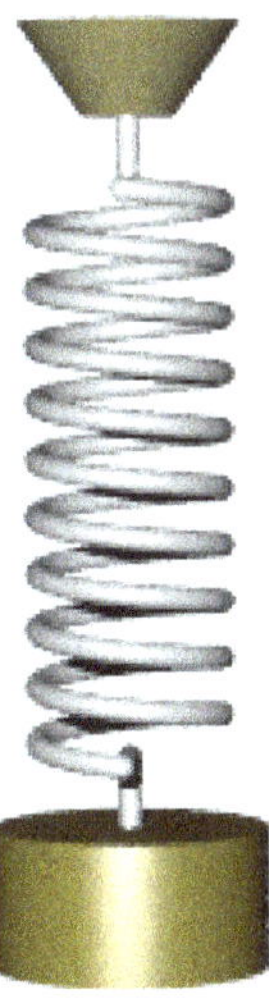

An undamped spring–mass system is an oscillatory system

Oscillations occur not only in mechanical systems but also in dynamic systems in virtually every area of science: for example the beating human heart, business cycles in economics, predator–prey population cycles in ecology, geothermal geysers in geology, vibrating strings in musical instruments, periodic firing of nerve cells in the brain, and the periodic swelling of Cepheid variable stars in astronomy.

Simple Harmonic Oscillator

The simplest mechanical oscillating system is a weight attached to a linear spring subject to only weight and tension. Such a system may be approximated on an air table or ice surface. The system is in an equilibrium state when the spring is static. If the system is displaced from the equilibrium, there is a net *restoring force* on the mass, tending to bring it back to equilibrium. However, in moving the mass back to the equilibrium position, it has acquired momentum which keeps it moving beyond that position, establishing a new restoring force in the opposite sense. If a constant force such as gravity is added to the system, the point of equilibrium is shifted. The time taken for an oscillation to occur is often referred to as the *oscillatory period*.

Systems where the restoring force on a body is directly proportional to its displacement, such as the dynamics of the spring-mass system, are described mathematically by the simple harmonic oscillator and the regular periodic motion is known as simple harmonic motion. In the spring-mass system, oscillations occur because, at the static equilibrium displacement, the mass has kinetic energy which is converted into potential energy stored in the spring at the extremes of its path. The spring-mass system illustrates some common features of oscillation, namely the existence of an equilibrium and the presence of a restoring force which grows stronger the further the system deviates from equilibrium.

Damped and Driven Oscillations

All real-world oscillator systems are thermodynamically irreversible. This means there are dissipative processes such as friction or electrical resistance which continually convert some of the energy stored in the oscillator into heat in the environment. This is called damping. Thus, oscillations tend to decay with time unless there is some net source of energy into the system. The simplest description of this decay process can be illustrated by oscillation decay of the harmonic oscillator.

In addition, an oscillating system may be subject to some external force, as when an AC circuit is connected to an outside power source. In this case the oscillation is said to be *driven*.

Some systems can be excited by energy transfer from the environment. This transfer typically occurs where systems are embedded in some fluid flow. For example, the phenomenon of flutter in aerodynamics occurs when an arbitrarily small displacement of an aircraft wing (from its equilibrium) results in an increase in the angle of attack of the wing on the air flow and a consequential increase in lift coefficient, leading to a still greater displacement. At sufficiently large displacements, the stiffness of the wing dominates to provide the restoring force that enables an oscillation.

Coupled Oscillations

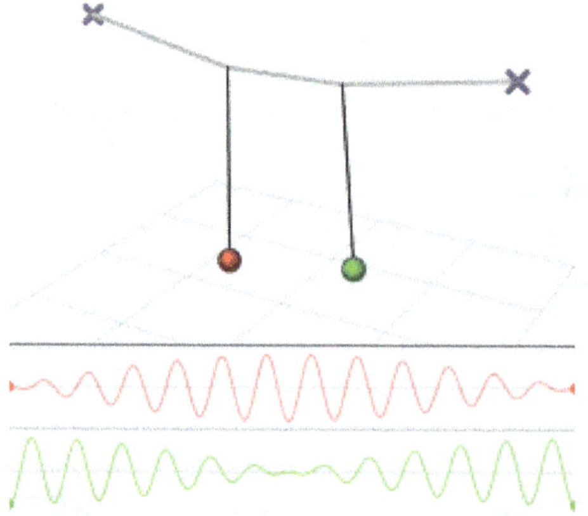

Two pendulums with the same period fixed on a string act as pair of coupled oscillators. The oscillation alternates between the two.

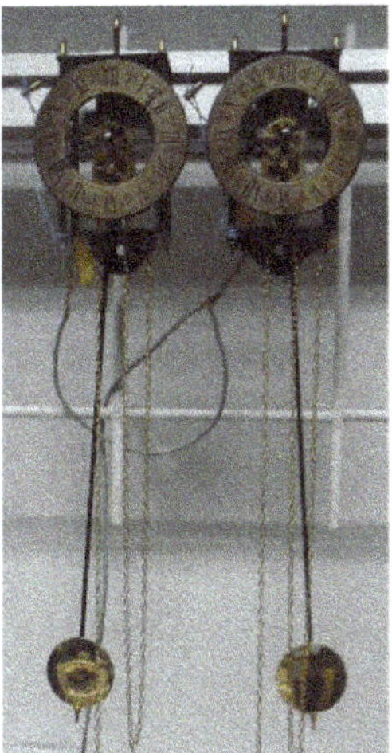

Experimental Setup of Huygens synchronization of two clocks

The harmonic oscillator and the systems it models have a single degree of freedom. More complicated systems have more degrees of freedom, for example two masses and three springs (each mass being attached to fixed points and to each other). In such cases, the behavior of each variable influences that of the others. This leads to a *coupling* of the oscillations of the individual degrees of freedom. For example, two pendulum clocks (of identical frequency) mounted on a common wall will tend to synchronise. This phenomenon was first observed by Christiaan Huygens in 1665. The apparent motions of the compound oscillations typically appears very complicated but a more economic, computationally simpler and conceptually deeper description is given by resolving the motion into normal modes.

More special cases are the coupled oscillators where energy alternates between two forms of oscillation. Well-known is the Wilberforce pendulum, where the oscillation alternates between an elongation of a vertical spring and the rotation of an object at the end of that spring.

Continuous Systems – Waves

As the number of degrees of freedom becomes arbitrarily large, a system approaches continuity; examples include a string or the surface of a body of water. Such systems have (in the classical limit) an infinite number of normal modes and their oscillations occur in the form of waves that can characteristically propagate.

Mathematics

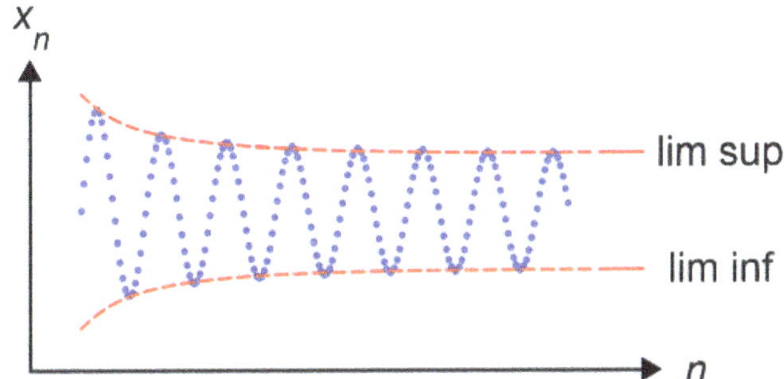

Oscillation of a sequence (shown in blue) is the difference between the limit superior and limit inferior of the sequence.

The mathematics of oscillation deals with the quantification of the amount that a sequence or function tends to move between extremes. There are several related notions: oscillation of a sequence of real numbers, oscillation of a real valued function at a point, and oscillation of a function on an interval (or open set).

Examples

Mechanical

- Double pendulum
- Foucault pendulum
- Helmholtz resonator
- Oscillations in the Sun (helioseismology), stars (asteroseismology) and Neutron-star oscillations.
- Quantum harmonic oscillator
- Playground swing
- String instruments
- Torsional vibration
- Tuning fork
- Vibrating string
- Wilberforce pendulum
- Lever escapement

Electrical

- Alternating current
- Armstrong (or Tickler or Meissner) oscillator
- Astable multivibrator
- Blocking oscillator
- Butler oscillator
- Clapp oscillator
- Colpitts oscillator
- Delay line oscillator
- Dow (or ultra-audion) oscillator

- Electronic oscillator
- Extended interaction oscillator
- Hartley oscillator
- Oscillistor
- Phase-shift oscillator
- Pierce oscillator
- Relaxation oscillator
- RLC circuit
- Royer oscillator
- Vačkář oscillator
- Wien bridge oscillator

Electro-mechanical

- Crystal oscillator

Optical

- Laser (oscillation of electromagnetic field with frequency of order 10^{15} Hz)
- Oscillator Toda or self-pulsation (pulsation of output power of laser at frequencies 10^4 Hz – 10^6 Hz in the transient regime)
- Quantum oscillator may refer to an optical local oscillator, as well as to a usual model in quantum optics.

Biological

- Circadian rhythm
- Circadian oscillator
- Lotka–Volterra equation
- Neural oscillation
- Oscillating gene
- Segmentation oscillator

Human

- Neural oscillation

- Insulin release oscillations
- gonadotropin releasing hormone pulsations
- Pilot-induced oscillation
- Voice production

Economic and Social

- Business cycle
- Generation gap
- Malthusian economics
- News cycle

Climate and Geophysics

- Atlantic multidecadal oscillation
- Chandler wobble
- Climate oscillation
- El Niño-Southern Oscillation
- Pacific decadal oscillation
- Quasi-biennial oscillation

Astrophysics

- Neutron stars
- Cyclic Model

Quantum Mechanical

- Neutrino oscillations
- Quantum harmonic oscillator

Chemical

- Belousov–Zhabotinsky reaction
- Mercury beating heart
- Briggs–Rauscher reaction
- Bray–Liebhafsky reaction

Computing

- Cellular Automata oscillator

Simple Harmonic Oscillators SHO

We consider the spring-mass system shown in Figure. A massless spring, one of whose ends is fixed has its other attached to a particle of mass m which is free to move. We choose the origin $x = 0$ for the particle's motion at the position where the spring is unstretched. The particle is in stable equilibrium at this position and it will continue to remain there if left at rest. We are interested in a situation where the particle is disturbed from equilibrium. The particle experiences a restoring force from the spring if it is either stretched or compressed. The spring is assumed to be elastic which means that it follows Hooke's law where the force is proportional to the displacement $F = -kx$ with spring constant k.

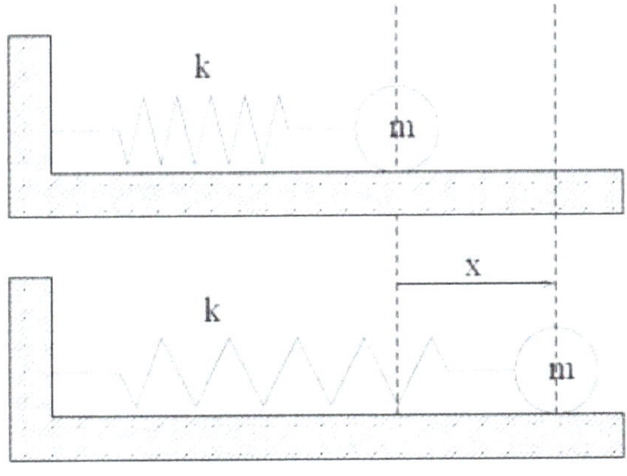

Spring-mass system

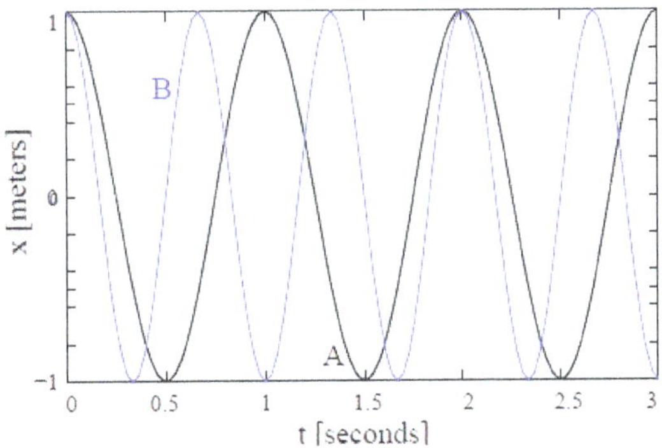

Displacement of the oscillator as a function of time for two different frequencies.

The particle's equation of motion is

$$m\frac{d^2x}{dt^2} = -kx$$

which can be written as $\ddot{x}+\omega_0^2 x=0$

where the dots denote time derivatives and

$$\omega_0=\sqrt{\frac{k}{m}}$$

It is straightforward to check that

$$x(t)=A\cos(\omega_0 t+\phi)$$

is a solution to eq.

We see that the particle performs sinusoidal oscillations around the equilibrium position when it is disturbed from equilibrium. The angular frequency ω_0 of the oscillation depends on the intrinsic properties of the oscillator. It determines the time period

$$T=\frac{2\pi}{\omega_0}$$

and the frequency $\nu=1/T$ of the oscillation. Figure shows oscillations for two different values of ω_0.

Problem 1: What are the values of ω_0 for the oscillations shown in Figure?

What are the corresponding spring constant k values if $m=1kg$?

Solution: For $A\ \omega_0=2\pi s^{-1}\ and\ k=(2\pi)^2 Nm^{-1};\ For\ B\ \omega_0=3\pi s^{-1}\ and\ k=(3\pi)^2 Nm^{-1}$

The amplitude A and phase ϕ are determined by the initial conditions.

Two initial conditions are needed to completely specify a solution. This follows from the fact that the governing equation is a second order differential equation. The initial conditions can be specified in a variety of ways, fixing the values of $x(t)\ and\ \dot{x}(t)\ at\ t=0$ is a possibility. Figure shows oscillations with different amplitudes and phases.

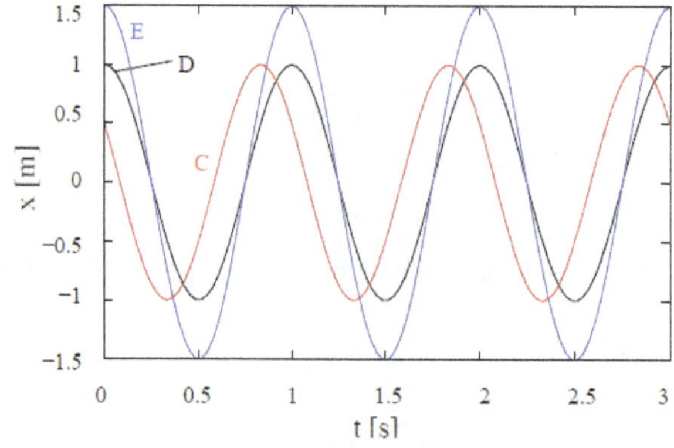

Displacement of the oscillator as a function of time for different initial conditions.

Problem 2: What are the amplitude and phase of the oscillations shown in Figure?

Solution: For $C, A=1\ and\ \phi=\pi/3; ForD, A=1\ and\ \phi=0; For\ E, A=1.5\ and\ \phi=0;$

Complex Representation.

Complex numbers provide are very useful in representing oscillations. The amplitude and phase of the oscillation can be combined into a single complex number which we shall refer to as the complex amplitude

$$\tilde{A}=Ae^{i\phi}.$$

Note that we have introduced the symbol ˜ (tilde) to denote complex numbers.

The property that

$$e^{i\phi}=\cos\phi+i\sin\phi$$

allows us to represent any oscillating quantity $x(t)=A\cos(\omega_0 t+\phi)$ as the real part of the complex number $\tilde{x}(t)=\tilde{A}e^{i\omega_0 t}$,

$$\tilde{x}(t)=Ae^{i(\omega_0 t+\phi)}=A\left[\cos(\omega_0 t+\phi)+i\sin(\omega_0 t+\phi)\right].$$

We calculate the velocity υ in the complex representation $\tilde{\upsilon}=\dot{\tilde{x}}$ which gives us

$$\tilde{\upsilon}(t)=i\omega_0\tilde{x}=-\omega_0 A\left[\sin(\omega_0 t+\phi)-i\cos(\omega_0 t+\phi)\right]$$

Taking only the real part we calculate the particle's velocity

$$\upsilon(t)=-\omega_0 A\sin(\omega_0 t+\phi).$$

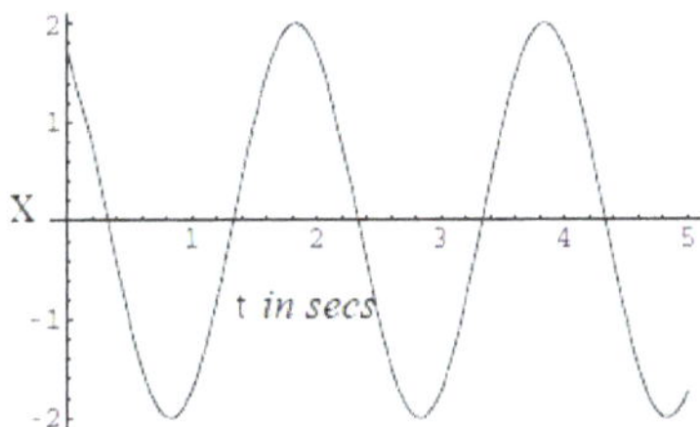

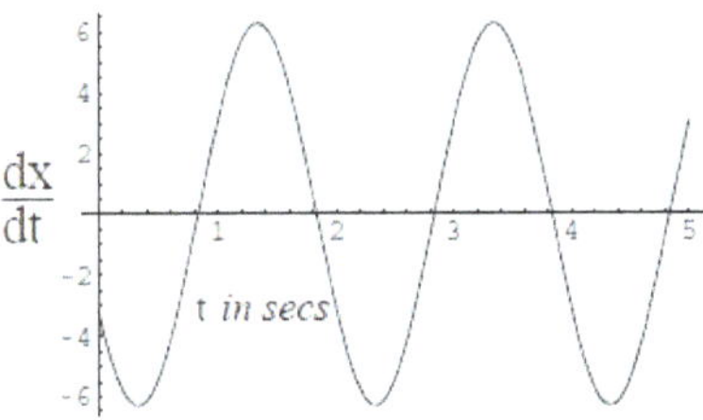

Displacement and velocity as a function of time.

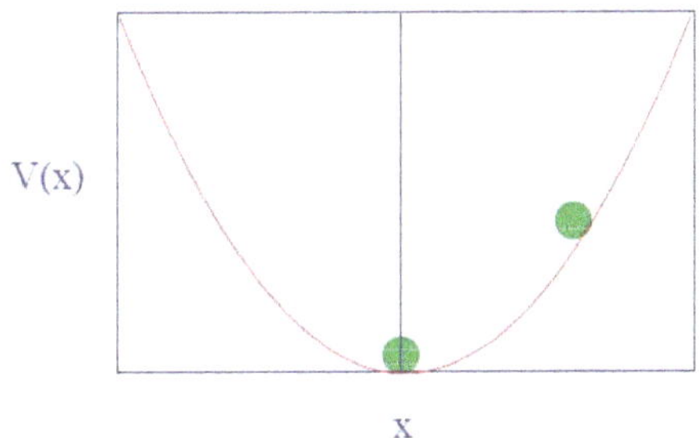

Harmonic oscillator potential energy.

The complex representation is a very powerful tool which, as we shall see later, allows us to deal with oscillating quantities in a very elegant fashion.

Figure shows the plots of displacement and velocity of the particle described by the equations, for amplitude, $A = 2$ units, phase, $\phi = 30°$ and angular frequency, $\omega_0 = \pi\ rad / sec$.

Problem 3: A SHO has position x_0 and velocity v_0 at the initial time $t = 0$. Calculate the complex amplitude $\tilde{A}$ in terms of the initial conditions and use this to determine the particle's position $x(t)$ at a later time t.

Solution: The initial conditions tell us that $\mathrm{Re}(\tilde{A}) = x_0$ *and* $\mathrm{Re}(i\omega_0\tilde{A}) = \upsilon_0$. Hence $\tilde{A} = x_0 - i\upsilon_0 / \omega_0$ which implies that $x(t) = x_0 \cos(\omega_0 t) + (\upsilon_0 / \omega_0) \sin(\omega_0 t)$.

Energy

In a spring-mass system the particle has a potential energy $V(x) = kx^2/2$ as shown in Figure. This energy is stored in the spring when it is either compressed or stretched. The potential energy of the system

$$U = \frac{1}{2}kA^2 \cos^2(\omega_0 t + \phi) = \frac{1}{4}m\omega_0^2 A^2 \{1 + \cos[2(\omega_0 t + \phi)]\}$$

Oscillates with angular frequency $2\omega_0$ as the spring is alternately compressed and stretched. The kinetic energy $m\upsilon^2/2$

$$T = \frac{1}{2}m\omega_0^2 \mathrm{A}^2 \sin^2(\omega_0 t + \phi) = \frac{1}{4}m\omega_0^2 A^2 \{1 - \cos[2(\omega_0 t + \phi)]\}$$

shows similar oscillations which are exactly π out of phase.

In a spring-mass system the total energy oscillates between the potential energy of the spring (U) and the kinetic energy of the mass (T). The total energy $E = T + U$ has a value $E = m\omega_0^2 A^2/2$ which remains constant.

The average value of an oscillating quantity is often of interest. We denote the time average of any quantity $Q(t)$ using $\langle Q \rangle$ which is defined as

$$\langle Q \rangle = \lim_{T\to\infty} \frac{1}{T} \int_{-T/2}^{T/2} Q(t)dt.$$

The basic idea here is to average over a time interval T which is significantly larger than the oscillation time period.

It is very useful to remember that $\langle \cos(\omega_0 t + \phi) \rangle = 0$. This can be easily verified by noting that the values $\sin(\omega_0 t + \phi)$ are bound between −1 and +1. We use this to calculate the average kinetic and potential energies both of which have the same values

$$\langle U \rangle = \langle T \rangle = \frac{1}{4}m\omega_0^2 A^2.$$

The average kinetic and potential energies, and the total energy are all very conveniently expressed in the complex representation as

$$E/2 = \langle U \rangle = \langle T \rangle = \frac{1}{4} = m\tilde{\upsilon}\tilde{\upsilon}^* = \frac{1}{4}k\tilde{x}\tilde{x}^*$$

where * denotes the conjugate of a complex number.

Problem 4: The mean displacement of a SHO $\langle x \rangle$ is zero. The root mean square (rms.) displacement $\sqrt{\langle x^2 \rangle}$ is useful in quantifying the amplitude of oscillation. Verify that the rms. displacement is $\sqrt{\tilde{x}\tilde{x}^*/2}$.

Solution: $\sqrt{\langle x^2(t) \rangle} = \sqrt{A^2 \langle \cos^2 |(\omega_0 t + \phi) \rangle} = \sqrt{A^2/2} = \sqrt{\tilde{A}e^{i\omega t}\tilde{A}^* e^{-i\omega t}/2} = \sqrt{\tilde{x}\tilde{x}^*/2}$

Why Study the SHO?

What happens to a system when it is disturbed from stable equilibrium? This question that arises in a large variety of situations. For example, the atoms in many solids (eg. NACl, diamond and steel) are arranged in a periodic crystal as shown in Figure. The periodic crystal is known to be an equilibrium configuration of the atoms. The atoms are continuously disturbed from their equilibrium positions as a consequence of random thermal motions and external forces which may happen to act on the solid. The study of oscillations in the atoms disturbed from their equilibrium position is very interesting. In fact the oscillations of the different atoms are coupled, and this gives rise to collective vibrations of the whole crystal which can explain properties like the specific heat capacity of the solid. We shall come back to this later, right now the crucial point is that each atom behaves like a SHO if we assume that all the other atoms remain fixed. This is generic to all systems which are slightly disturbed from stable equilibrium.

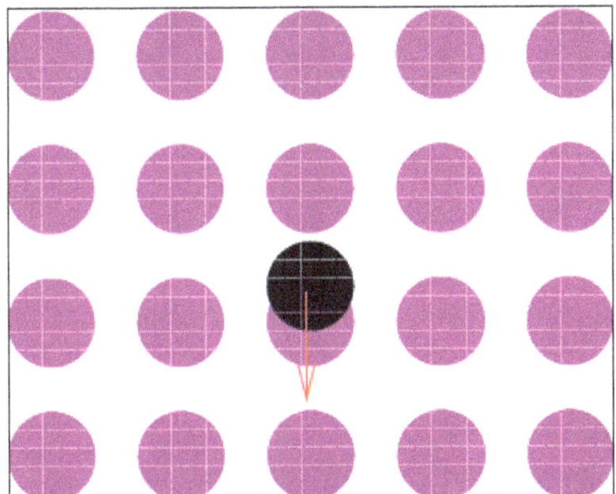

Atoms in a Crystal.

We now show that any potential $V(x)$ is well represented by a SHO potential in the neighbourhood of points of stable equilibrium. The origin of x is chosen so that the point of stable equilibrium is located at $x = 0$. For small values of x it is possible to approximate the function $V(x)$ using a Taylor series

$$V(x) \approx V(x)_{x=0} + \left(\frac{dV(x)}{dx}\right)_{x=0} x + \frac{1}{2}\left(\frac{d^2V(x)}{dx^2}\right)_{x=0} x^2 + \ldots$$

where the higher powers of x are assumed to be negligibly small. We know that at the points of stable equilibrium the force vanishes ie. $F = -dV(x)/dx = 0$ and $V(x)$ has a minimum.

$$k = \left(\frac{d^2V(x)}{dx^2}\right)_{x=0} > 0.$$

This tells us that the potential is approximately

$$V(x) \approx V(x)_{x=0} + \frac{1}{2}kx^2$$

which is a SHO potential. Figure shows two different potentials which are well approximated by the same SHO potential in the neighbourhood of the point of stable equilibrium. The oscillation frequency is exactly the same for particles slightly disturbed from equilibrium in these three different potentials.

The study of SHO is important because it occurs in a large variety of situations where the system is slightly disturbed from equilibrium. We discuss a few simple situations.

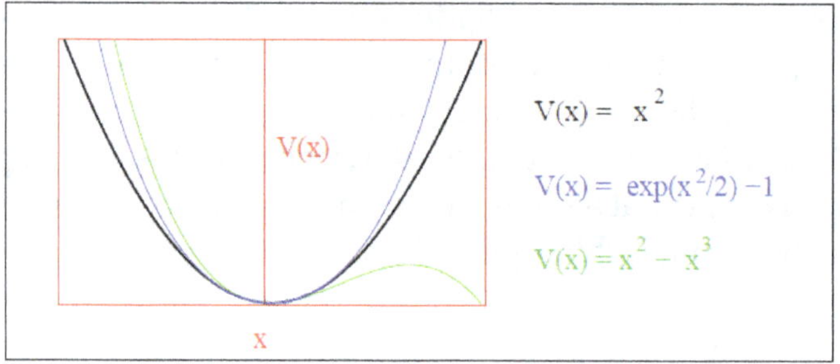

Various Potentials.

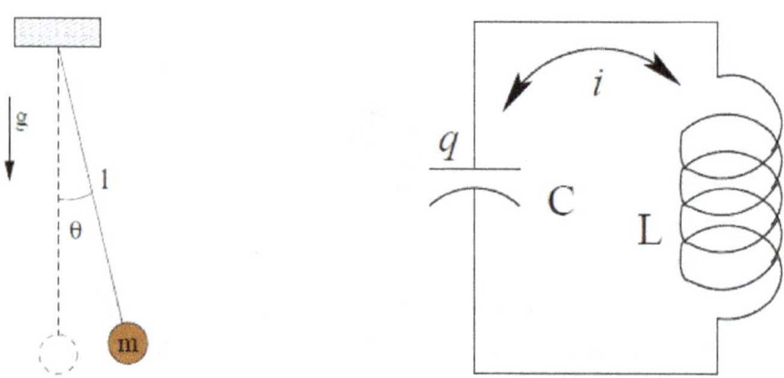

(a) Simple Pendulum and (b) LC Circuit.

Simple Pendulum

The simple pendulum shown in Figure (a) is possibly familiar to all of us. A mass m is suspended by a rigid rod of length l, the rod is assumed to be massless. The gravitations potential energy of the mass is

$$V(\theta) = mgl[1 - \cos\theta].$$

For small θ we may approximate $\cos\theta \approx 1-\theta^2/2$ whereby the potential is

$$V(\theta)=\frac{1}{2}mgl\theta^2$$

which is the SHO potential. Here $dV(\theta)/d\theta$ gives the torque not the force. The pendulum's equation of motion is

$$I\ddot{\theta}=-mgl\theta$$

where $I = ml^2$ is the moment of inertia. This can be written as

$$\ddot{\theta}+\frac{g}{l}\theta=0$$

which allows us to determine the angular frequency

$$\omega_0=\sqrt{\frac{g}{l}}$$

LC Oscillator

The LC circuit shown in Figure (b) is an example of an electrical circuit which is a SHO. It is governed by the equation

$$L\dot{I}+\frac{Q}{C}=0$$

where L refers to the inductance, C capacitance, I current and Q charge. This can be written as

$$\ddot{Q}+\frac{1}{LC}Q=0$$

which allows us to identify

$$\omega_0=\sqrt{\frac{1}{LC}}$$

as the angular frequency.

Torsional Pendulum

The equation for the torsional pendulum is the following.

$$I\frac{d^2\theta}{dt^2}+\kappa\theta=0,$$

where I is the moment of inertia of the object undergoing torsional oscillation about the axis of rotation and k is the torsional constant. Angular frequency

can be read off directly as $\omega_0 = \sqrt{\frac{\kappa}{I}}$ and hence the time period, $T = 2\pi\sqrt{\frac{I}{k}}$.

Physical Pendulum or Compound Pendulum

The equation of motion for a compound pendulum shown in is,

$$I\frac{d^2\theta}{dt^2} = -M\,gd\sin\theta,$$

where I is the moment of inertia about an axis perpendicular to the plane of oscillations through the point of suspension. For small oscillations $(\theta < 4^0)$ one can write the above equation approximately as,

$$\frac{d^2\theta}{dt^2} + \frac{Mgd}{I}\theta = 0.$$

The above gives time period as $T = 2\pi\sqrt{\frac{I}{Mgd}}$.

Problem 5: Obtain the simple pendulum results as a special case of the compound pendulum.

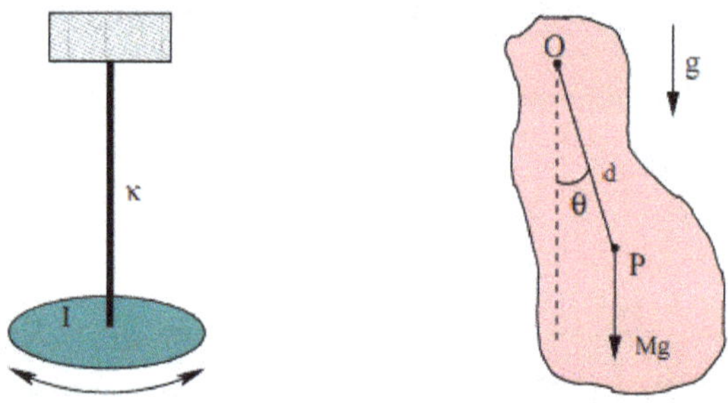

(a) Torsional pendulum and (b) Physical pendulum.

Damped Oscillator

Damping usually comes into play whenever we consider motion. We study the effect of damping on the spring-mass system. The damping force, will assumed to be proportional to the velocity, is acting to oppose the motion. The total force acting on the mass is

$$F = -kx - c\dot{x}$$

where in addition to the restoring force $-kx$ due to the spring we also have the damping force $-c\dot{x}$. The equation of motion for the damped spring mass system is

$$m\ddot{x} = -kx - c\dot{x}.$$

Recasting this in terms of more convenient coefficients, we have

$$\ddot{x}+2\beta\dot{x}+\omega_0^2 x=0$$

This is a second order homogeneous equation with constant coefficients. Both ω_0 *and* $\beta=c/(2m)$ have dimensions (time)$^{-1}$. Here $1/\omega_0$ is the time-scale of the oscillations that would occur if there was no damping, and $1/\beta$ is the time-scale required for damping to bring any motion to rest. It is clear that the nature of the motion depends on which time-scale $1/\omega_0$ *or* $1/\beta$ is larger.

We proceed to solve equation by taking a trial solution

$$x(t)=Ae^{\alpha t}.$$

Putting the trial solution into equation gives us the quadratic equation

$$\alpha^2+2\beta\alpha+\omega_0^2=0$$

This has two solutions

$$\alpha_1=-\beta+\sqrt{\beta^2-\omega_0^2}$$

And

$$\alpha_2=-\beta-\sqrt{\beta^2-\omega_0^2}$$

The nature of the solution depends critically on the value of the damping Coefficient β, and the behaviour is quite different depending on whether $\beta<\omega_0$, $\beta=\omega_0$ *or* $\beta>\omega_0$.

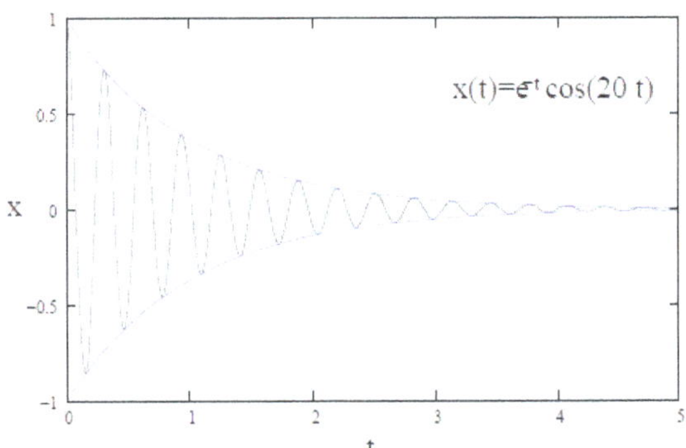

Underdamped Oscillations

We first consider the situation where $\beta<\omega_0$ which is referred to as underdamped. Defining

$$\omega=\sqrt{\omega_0^2-\beta^2}$$

the two roots which are both complex have values

$$\alpha_1=-\beta+i\omega \textit{ and } \alpha_2=-\beta-i\omega$$

The resulting solution is a superposition of the two roots

$$x(t) = e^{-\beta t}\left[A_1 e^{i\omega t} + A_2 e^{-i\omega t}\right]$$

where A1 and A2 are constants which have to be determined from the initial conditions. The term $\left[A_1 e^{i\omega t} + A_1 e^{-i\omega t}\right]$ is a superposition of sin and cos which can be written as

$$x(t) = Ae^{-\beta t}\cos(\omega t + \phi)$$

This can also be expressed in the complex notation as

$$\tilde{x}(t) = \tilde{\mathrm{A}}\mathrm{e}^{(i\omega - \beta)t}$$

Where $\tilde{A}=Ae^{i\phi}$ is the complex amplitude which has both the amplitude and phase information. Figure shows the underdamped motion $x(t) = e^{-1}\cos(20t)$, whereas Figure (a) shows the underdamped motion $x(t) = \sqrt{2}\exp\left(\frac{-t}{20}\right)\cos\left(\frac{t-\pi}{4}\right)$

In all cases damping reduces the frequency of the oscillations i.e. $\omega < \omega_0$. The main effect of damping is that it causes the amplitude of the oscillations to decay exponentially with time. It is often useful to quantify the decay in the amplitude during the time period of a single oscillation $T = 2\pi/\omega$. This is quantified by the logarithmic decrement which is defined as

$$\lambda = ln\left[\frac{x(t)}{x(t+T)}\right] = \frac{2\pi\beta}{\omega}$$

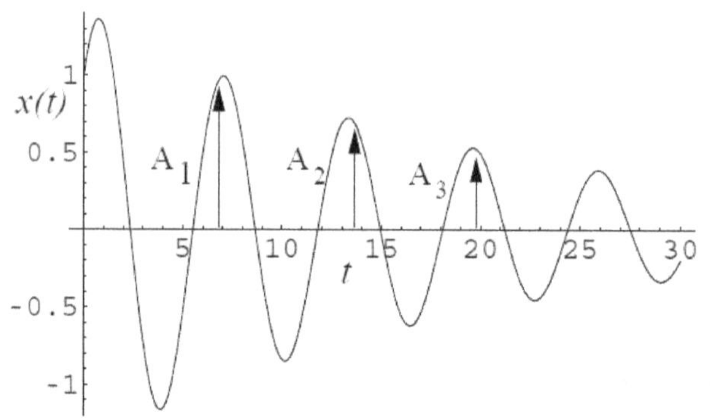

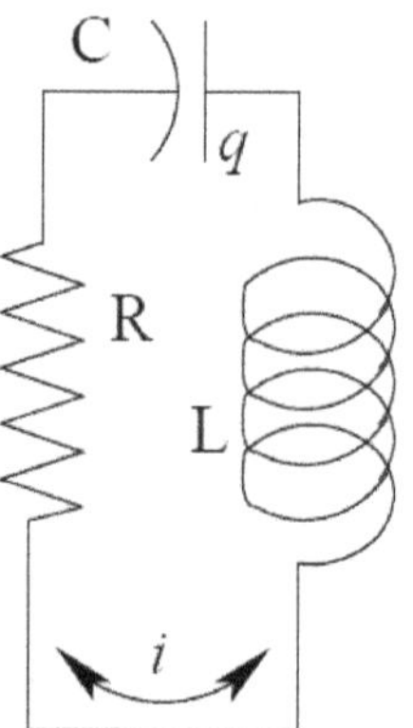

(a) and (b)

From figure (a) it is clear that $\lambda = ln = \left(\frac{A_n}{A_{n+1}}\right)$

Energy stored in a damped harmonic oscillator In the case of a damped oscillator the total energy of the system decreases with time. Since a damping factor $\exp(-\beta t)$ is present in the expression of displacement $x(t)$, the total energy is given by,

$$E(t) = \frac{1}{2}\exp(-2\beta t)kA^2$$

where A is the initial amplitude.

Example: Damped SHM in LCR circuit: The voltage equation for the circuit is,

$$L\frac{di}{dt} + Ri + q/C = 0$$
$$or,\ L\ddot{q} + R\dot{q} + q/C = 0.$$

where i is the current in the circuit and q is the charge stored in the capacitor. Comparing with equation we obtain the solution c for charge on the capacitor as,

$$q(t) = \exp(-Rt/2L)\left[q_+ \exp\sqrt{(R^2/4L^2)-(1/LC)}t\right]$$

$$+q\_\exp\left[-\sqrt{(R^2/4L^2)-(1/LC)}t\right]$$

where q_+ & q_- are determined from the initial condition.

Over-Damped Oscillations

This refers to the situation where

$$\beta > \omega_0$$

The two roots are

$$\alpha_1 = -\beta + \sqrt{\beta^2 - \omega_0^2} = -\gamma_1$$

And

$$\alpha_2 = -\beta - \sqrt{\beta^2 - \omega_0^2} = -\gamma_2$$

where both $\gamma_1, \gamma_2 > 0$ and $\gamma_2 > \gamma_1$. The two roots give rise to exponentially decaying solutions, one of which decays faster than the other

$$x(t) = A_1 e^{-\gamma_1 t} + A_2 e^{-\gamma_2 t}.$$

The constants A_1 and A_2 are determined by the initial conditions. For initial position x_0 and velocity u_0 we have

$$x(t) = \frac{\upsilon_0 + \gamma_2 x_0}{\gamma_2 - \gamma_1} e^{-\gamma_1 t} - \frac{\upsilon_0 + \gamma_1 x_0}{\gamma_2 - \gamma_1} e^{-\gamma_2 t}$$

The overdamped oscillator does not oscillate. Figure shows a typical situation.

In the situation where $\beta >> \omega_0$

$$\sqrt{\beta^2 - \omega_0^2} = \beta\sqrt{1 - \frac{\omega_0^2}{\beta^2}} \approx \beta\left[1 - \frac{1}{2}\frac{\omega_0^2}{\beta^2}\right]$$

and we have $\gamma_1 = \omega_0^2 / 2\beta$ *and* $\gamma_2 = 2\beta$.

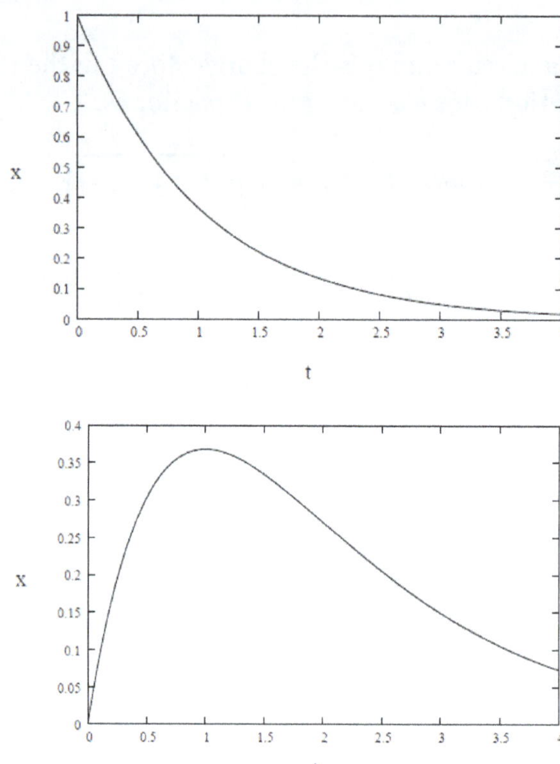

Critical Damping

This corresponds to a situation where $\beta = \omega_0$ and the two roots are equal. The governing equation is second order and there still are two independent solutions. The general solution is

$$x(t) = e^{-\beta t}\left[A_1 + A_2 t\right]$$

The solution

$$x(t) = x_0 e^{-\beta t}\left[1 + \beta t\right]$$

is for an oscillator starting from rest at x_0 while

$$x(t) = \upsilon_0 e^{-\beta t} t$$

is for a particle starting from $x = 0$ with speed υ_0.

Figure shows a typical comparision of the three types of damping viz, underdamped, overdamped and critically damped. Figure shows the comparision of a critically damped oscillator with an over damped oscillator for different values of β. One observes that the critically damped oscillator reaches the mean position in the smallest possible time. This is the reason that the resistances in seat-shock absorber of vehicles, in sliding doors or in Galvanometers are adjusted to the critically damped condition so that when they are disturbed they come back to the mean position quickly.

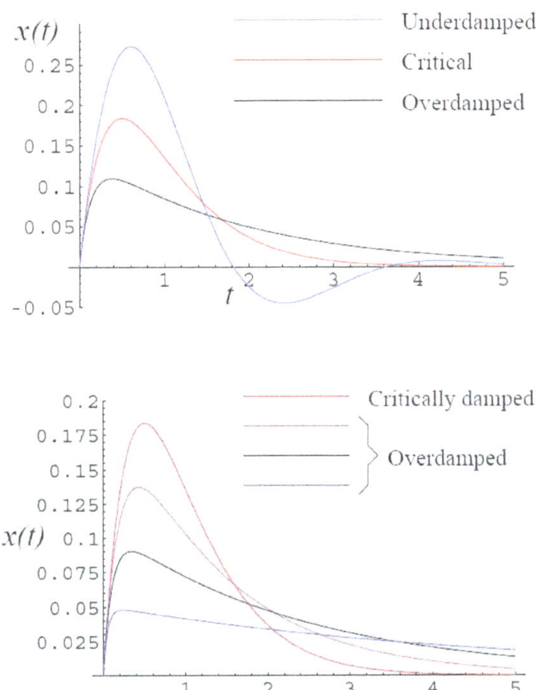

Oscillator with External Forcing

In this chapter we consider an oscillator under the influence of an external sinusoidal force $F = \cos(\omega t + \psi)$. Why this particular form of the force? This is because nearly any arbitrary time varying force $F(t)$ can be decomposed into the sum of sinusoidal forces of different frequencies

$$F(t) = \sum_{n=1,\ldots}^{\infty} F_n \cos(\omega_n t + \psi_n)$$

Here F_n and ψ_n are respectively the amplitude and phase of the different frequency components. Such an expansion is called a Fourier series. The behaviour of the oscillator under the influence of the force $F(t)$ can be determined by separately solving

$$m\ddot{x}_n + kx_n = F_n \cos(\omega_n t + \psi_n)$$

for a force with a single frequency and then superposing the solutions

$$x(t) = \sum_n x_n(t).$$

We shall henceforth restrict our attention to equation which has a sinusoidal force of a single frequency and drop the subscript n from x_n and F_n. It is convenient to switch over to the complex notation

$$\ddot{\tilde{x}} + \omega_0^2 \tilde{x} = \tilde{f} e^{i\omega t}$$

Where $\tilde{f} = Fe^{i\psi} / m.$

Complementary Function and Particular Integral

The solution is a sum of two parts

$$\tilde{x}(t) = \tilde{A} e^{i\omega_0 t} + \tilde{B} e^{i\omega t}.$$

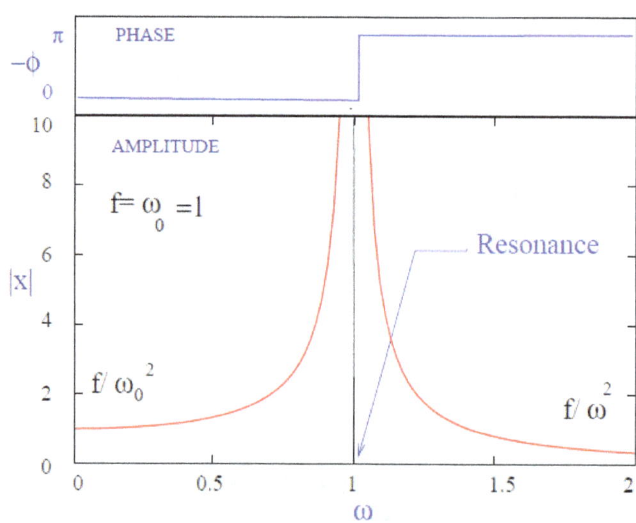

Amplitude and phase as a function of forcing frequency

The first term $\tilde{A} e^{i\omega_0 t}$, called the complementary function, is a solution to equation without the external force. This oscillates at the natural frequency of the oscillator ω_0. This part of the solution is exactly the same as when there is no external force. This has been discussed extensively earlier, and we shall ignore this term in the rest of this chapter.

The second term $Be^{i\ t}$, called the particular integral, is the extra ingredient in the solution due to the external force. This oscillates at the frequency of the external force ω. The amplitude $\tilde{B}$ is determined from equation which gives

$$\left[-\omega^2 + \omega_0^2\right] \tilde{B} = \tilde{f}$$

Whereby we have the solution

$$\tilde{x}(t)=\frac{\tilde{f}}{\omega_0^2-\omega^2}e^{i\omega t}.$$

The amplitude and phase of the oscillation both depend on the forcing frequency ω. The amplitude is

$$|\tilde{x}|=\frac{f}{|\omega_0^2-\omega^2|}.$$

and the phase of the oscillations relative to the applied force is ϕ =0 for $\omega<\omega_0$ *and* $\phi=-\pi$ *for* $\omega>\omega_0$.

Note: One cannot decide here whether the oscillations lag or lead the driving force, i.e. whether $\phi=-\pi$ *or* $\phi=\pi$ as both of them are consistent with $\omega>\omega_0$ *case*$(e^{\pm i\pi}=-1)$. The zero resistance limit, $\beta\to 0$ of the damped forced oscillations would settle it for $\phi=-\pi$ *for* $\omega>\omega_0$. So in this case there is an abrupt change of $-\pi$ radians in the phase as the forcing frequency, ω, crosses the natural frequency, ω_0.

The amplitude and phase are shown in Figure. The first point to note is that the amplitude increases dramatically as $\omega\to\omega_0$ and the amplitude blows up at $\omega=\omega_0$. This is the phenomenon of resonance. The response of the oscillator is maximum when the frequency of the external force matches the natural frequency of the oscillator. In a real situation the amplitude is regulated by the presence of damping which ensures that it does not blow up to infinity at $\omega=\omega_0$.

We next consider the low frequency $\omega\ll\omega_0$ behaviour

$$\tilde{x}(t)=\frac{\tilde{f}}{\omega_0^2}e^{i\omega t}=\frac{F}{k}e^{i(\omega t+\phi)},$$

The oscillations have an amplitude F/k and are in phase with the external force.

This behaviour is easy to understand if we consider $\omega=0$ which is a constant force. We know that the spring gets extended (or contracted) by an amount $x=F/k$ in the direction of the force. The same behaviour goes through if F varies very slowly with time. The behaviour is solely determined by the spring constant k and this is referred to as the "Stiffness Controlled" regime.

At high frequencies $\omega\gg\omega_0$

$$\tilde{x}(t)=-\frac{\tilde{f}}{\omega^2}e^{i\omega t}=-\frac{F}{m\omega^2}e^{i(\omega t+\phi)},$$

the amplitude is F/m and the oscillations are $-\pi$ out of phase with respect to the force. This is the "Mass Controlled" regime where the spring does not come into the picture at all. It is straight forward to verify that equation is a solution to

$$m\ddot{x}=Fe^{i(\omega t+\phi)}$$

when the spring is removed from the oscillator. Interestingly such a particle moves exactly out of phase relative to the applied force. The particle moves to the left when the force acts to the right and vice versa.

Undamped Forced Oscillations and Resonance

We shall now assume a sinusoidal time dependence $(F(t) = F_0 \sin \omega t))$ of forcing. The angular frequency ω appearing in the driving force is called the *driving frequency*. Why we would like to study the sinusoidal forcing will become clear as we proceed. So the equation we are interested in solving, is

$$\ddot{x} + \omega_0^2 x = \frac{F_0}{m} \sin \omega t = f_0 \sin \omega t,$$

where, $F_0 / m = f_0$. We already know the complementary function for the above equation. The particular solution is given by

$$\text{A}=\frac{f_0}{\left(\omega_0^2 - \omega^2\right)} \sin \omega t.$$

The general solution can now be written as by adding the complementary function

$$\text{x}(t)=\frac{f_0}{\omega_0^2 - \omega^2} \sin \omega t + B \cos \omega_0 t + C \sin \omega_0 t$$

We fix B and C using some initial conditions. Let us choose $x(t=0) = \dot{x}(t=0) = 0$. The condition $x(0) = 0$ *fixes* $B = 0$ *and* $\dot{x}(0)$ *finds* $C = -\frac{f_0 \omega}{\omega_0\left(\omega_0^2 - \omega^2\right)} = -A\omega / \omega_0$. Hence the solution becomes,

$$x(t) = A(\sin \omega t - \frac{\omega}{\omega_0} \sin \omega_0 t$$

Figure is a sample plot of behaviour away from resonance,

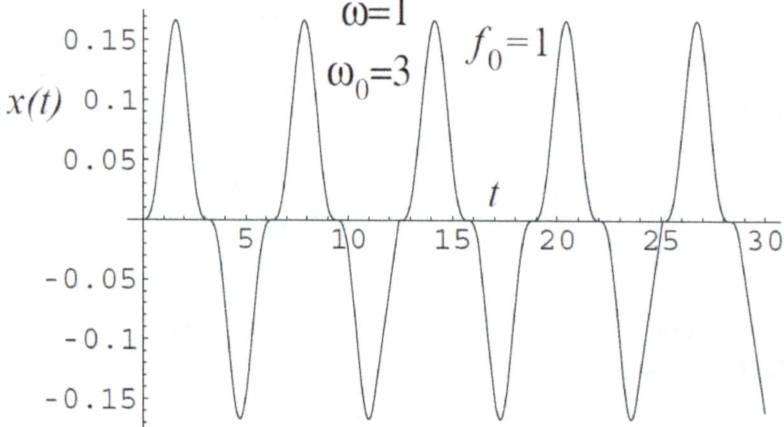

Undamped forced oscillations away from resonance

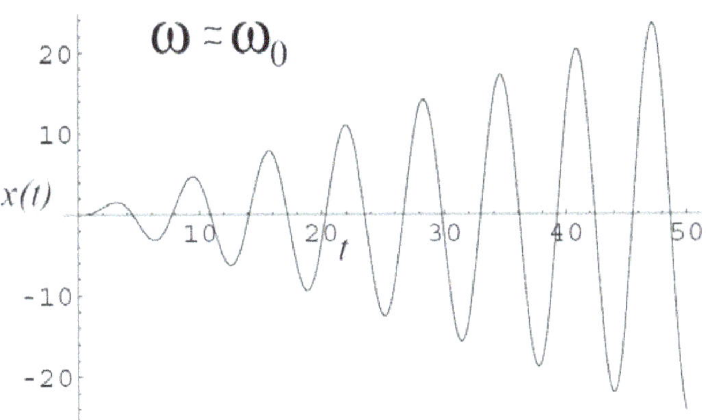

Undamped forced oscillations, behaviour near resonance

We would now like to investigate the behaviour of general solution

near resonance. Let us take $\omega = \omega_0 - \Delta\omega$,

$$x(t) = A\left(\sin\omega_0 t\cos\Delta\omega t - \cos\omega_0 t\sin\Delta\omega t - \frac{\omega}{\omega_0}\sin\omega_0 t\right)$$

$$x(t) = A\left(\frac{(\omega_0 - \omega)}{\omega_0}\sin\omega_0 t - \Delta\omega t\cos\omega_0 t\right)$$

where in equation we have used $\cos\Delta\omega t \approx 1$ *and* $\sin\Delta\omega t \approx \Delta\omega t$. Substituting the value of A from we get,

$$x(t) = \frac{f_0}{\omega_0(\omega_0 + \omega)}(\sin\omega_0 t - \omega_0 t\cos\omega_0 t)$$

$$\approx \frac{f_0}{2\omega_0^2}(\sin\omega_0 t - \omega_0 t\cos\omega_0 t)$$

The second term in equation grows with time making the amplitude grow as well. Figure shows the behaviour of the undamped forced oscillator near the resonance. One sees that the amplitude grows indefinitely and the oscillator reaches a point where it cannot sustain any further oscillations and it breaks.

Effect of Damping

Introducing damping, the equation of motion

$$m\ddot{x} + c\dot{x} + kx = F\cos(\omega t + \psi)$$

written using the notation introduced earlier is

$$\ddot{\tilde{x}} + 2\beta\dot{\tilde{x}} + \omega_0^2\tilde{x} = \tilde{f}e^{i\omega t}.$$

Here again we separately discuss the complementary functions and the particular integral. The complementary functions are the decaying solutions that arise when there is no external force. These are short lived transients which are not of interest when studying the long time behaviour of the oscillations. These have already been discussed in considerable detail and we do not consider them here. The particular integral is important when studying the long time or steady state response of the oscillator. This solution is

$$\tilde{x}(t) = \frac{\tilde{f}}{(\omega_0^2 - \omega^2) + 2i\beta\omega} e^{i\omega t}$$

which may be written as $\tilde{x}(t) = Ce^{i\phi}\tilde{f}e^{i\omega t}$ where ϕ is the phase of the oscillation relative to the force $\tilde{f}$.

This has an amplitude

$$|\tilde{x}| = \frac{f}{\sqrt{(\omega_0^2 - \omega^2)^2 + 4\beta^2\omega^2}}$$

and the phase ϕ is

$$\phi = \tan^{-1}\left(\frac{-2\beta\omega}{\omega_0^2 - \omega^2}\right)$$

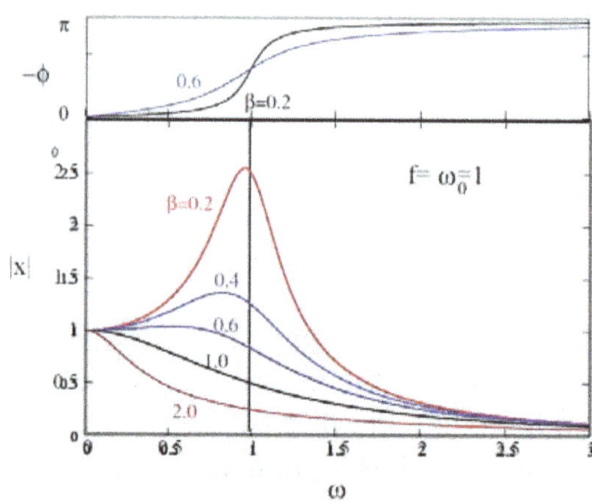

Amplitudes and phases for various damping coefficients as a function of driving frequency

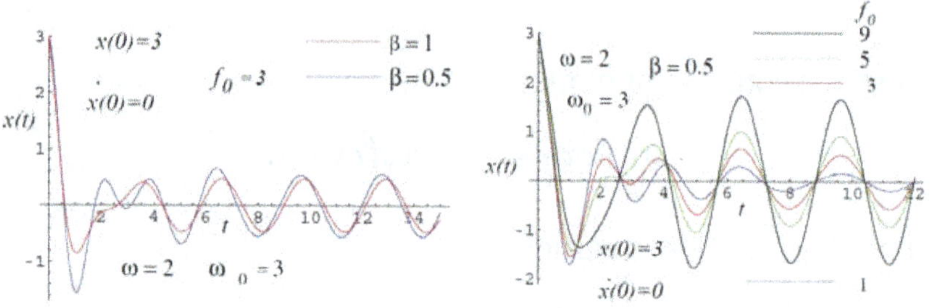

Forced oscillations with different resistances and forcing amplitudes

Figure shows the amplitude and phase as a function of ω for different values of the damping coefficient β. The damping ensures that the amplitude does not blow up at $\omega = \omega_0$ and it is finite for all values of ω. The change in the phase also is more gradual.

Problem 1: Plot the response, $x(t)$, of a forced oscillator with a forcing 3 $cos\ 2t$ and natural frequency ω_0 = 3 Hz with initial conditions, $x(0)$ = 3 and $\dot{x}(0)$ = 0, for two different resistances, β = 1 and β = 0.5. Plot also for fixed resistance, β = 0.5 and different forcing amplitudes $f_0 = 1,3,5$ *and* 9.

Solution : The evolution is shown in the Figure. Notice that the transients die and the steady state is achieved relatively sooner in the case of larger resistance, $\beta = 1$. Furthermore, the steady state is reached quicker in the case of larger forcing amplitude.

The low frequency and high frequency behaviour are exactly the same as the situation without damping. The changes due to damping are mainly in the vicinity of $\omega = \omega_0$. The amplitude is maximum at

$$\omega = \sqrt{\omega_0^2 - 2\beta^2}$$

For mild damping ($\beta \ll \omega_0$) this is approximately $\omega = \omega_0$.

We next shift our attention to the energy of the oscillator. The average energy $E(\omega)$ is the quantity of interest. Calculating this as a function of ω we have

$$E(\omega) = \frac{mf^2}{4} \frac{\left(\omega^2 + \omega_0^2\right)}{\left[\left(\omega_0^2 - \omega^2\right)^2 + 4\beta^2\omega^2\right]}$$

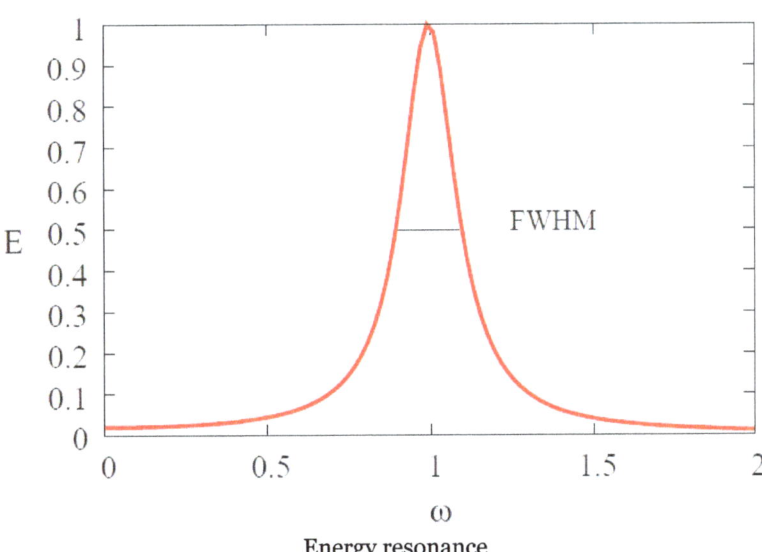

Energy resonance

The response to the external force shows a prominent peak or resonance only when $\beta \ll \omega_0$, the mild damping limit. This is of great utility in modelling the phenomena of resonance which occurs in a large variety of situations. In the weak damping limit $E(\omega)$ peaks at $\omega \approx \omega_0$ and falls rapidly away from the peak. As a consequence we can use

$$\left(\omega_0^2-\omega^2\right)^2=\left(\omega_0+\omega\right)^2\left(\omega_0-\omega\right)^2 \approx 4\omega_0^2\left(\omega_0-\omega\right)^2$$

which gives

$$E(\omega) \approx \frac{k}{8}\frac{f^2}{\omega_0^2\left[\left(\omega_0-\omega\right)^2+\beta^2\right]}$$

in the vicinity of the resonance. This has a maxima at $\omega \approx \omega_0$ and the maximum value is

$$E_{\max} \approx \frac{kf^2}{8\omega_0^2\beta^2}.$$

We next estimate the width of the peak or resonance. This is quantified using the FWHM (Full Width at Half Maxima) defined as $FWHM = 2\Delta\omega$ where $E(\omega_0+\Delta\omega)=E_{max}/2$ ie. half the maximum value. Using equation we see that $\Delta\omega=\beta$ and $FWHM=2\beta$. As shown in Figure. The FWHM quantifies the width of the curve and it records the fact that the width increases with the damping coefficient β.

The peak described by equation is referred to as a Lorentzian profile. This is seen in a large variety of situations where we have a resonance.For example, Intensity profiles of spectral lines are Lorentzian. We finally consider the power drawn by the oscillator from the external force. The instantaneous power $P(t)=F(t)\dot{x}(t)$ has a value

$$P(t)=\left[F\cos(\omega t)\right]\left[-|\tilde{x}|\omega\sin(\omega t+\phi)\right].$$

The average power is the quantity of interest, we study this as a function of the frequency. Calculating this we have

$$\langle P\rangle(w)=-\frac{1}{2}\omega F|\tilde{x}|\sin\phi.$$

Using equation we have

$$|\tilde{x}|\sin\phi=\frac{-2\beta\omega}{\left(\omega_0^2-\omega^2\right)^2+4\beta^2\omega^2}\left(\frac{F}{m}\right)$$

which gives the average power

$$\langle P\rangle(\omega)=\frac{\beta\omega^2}{\left(\omega_0^2-\omega^2\right)^2+4\beta^2\omega^2}\left(\frac{F^2}{m}\right)$$

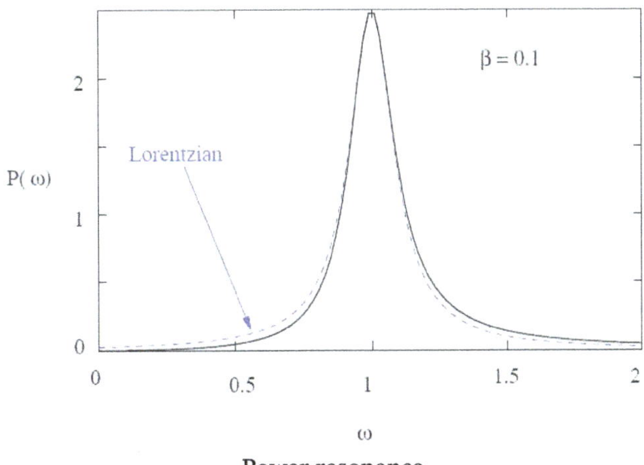

Power resonance

The solid curve in Figure shows the average power as a function of ω. Here again, a prominent, sharp peak is seen only if $\beta \ll \omega_0$. In the mild damping limit, in the vicinity of the maxima we have

$$\langle P \rangle(\omega) \approx \frac{\beta}{(\omega_0 - \omega)^2 + \beta^2}\left(\frac{F^2}{4m}\right).$$

which again is a Lorentzian profile. For comparison we have also plotted the Lorentzian profile as a dashed curve in Figure.

Problem 2: The galvanometer: A galvanometer is connected with a constant-current source through a switch. At time $t = 0$, the switch is closed. After some time the galvanometer deflection reaches its final value θ_{max}. Taking damping torque proportional to the angular velocity draw the deflection of the galvanometer from the initial position of rest $(i.e. \theta = 0, \theta = 0)$ to its final position $\theta = \theta_{max}$, for the underdamped, critically damped and overdamped cases.

Solution: We solve the forced oscillator equation with constant forcing (i.e. driving frequency =0) and given initial conditions and plot the various evolutions. Figure shows the galvanometer deflection as a function of time for some arbitrary values of θ_{max}, damping coefficient and natural frequency.

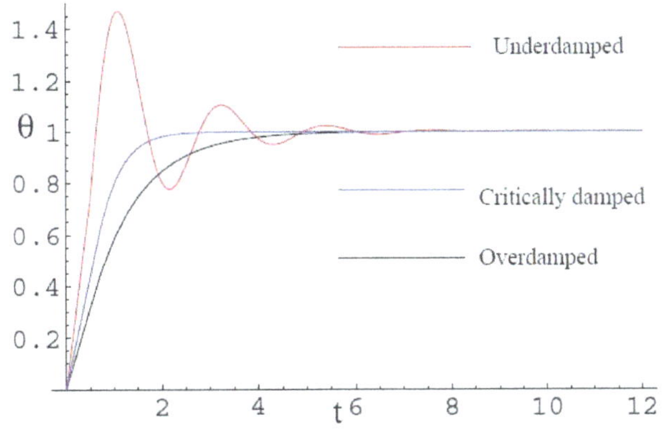

Galvanometer deflection

2

Resonance: An Integrated Study

The concept by which a system oscillates with greater amplitude and at specific frequencies with the application of a vibrating system or an external force is known as resonance. Some of the examples of resonance are swings, Tacoma narrows bridge, modern watches, etc. The topics discussed in this chapter are coupled oscillators, wave, sine wave and electrical circuits. The diverse applications of resonance in the current scenario have been thoroughly discussed in this chapter.

Resonance

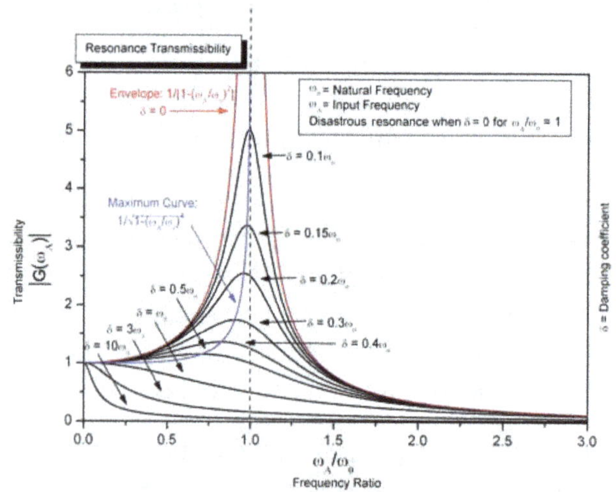

Increase of amplitude as damping decreases and frequency approaches resonant frequencyof a driven damped simple harmonic oscillator.

In physics, resonance is a phenomenon in which a vibrating system or external force drives another system to oscillate with greater amplitude at specific frequencies.

Frequencies at which the response amplitude is a relative maximum are known as the system's resonant frequencies or resonance frequencies. At resonant frequencies, small periodic driving forces have the ability to produce large amplitude oscillations, due to the storage of vibrational energy.

Overview

Resonance occurs when a system is able to store and easily transfer energy between two or more different storage modes (such as kinetic energy and potential energy in the case of a pendulum). However, there are some losses from cycle to cycle, called damping. When damping is small, the resonant frequency is approximately equal to the natural frequency of the system, which is a frequency of unforced vibrations. Some systems have multiple, distinct, resonant frequencies.

Resonance phenomena occur with all types of vibrations or waves: there is mechanical resonance, acoustic resonance, electromagnetic resonance, nuclear magnetic resonance (NMR), electron spin resonance (ESR) and resonance of quantum wave functions. Resonant systems can be used to generate vibrations of a specific frequency (e.g., musical instruments), or pick out specific frequencies from a complex vibration containing many frequencies (e.g., filters).

The term *resonance* originates from the field of acoustics, particularly observed in musical instruments, e.g., when strings started to vibrate and to produce sound without direct excitation by the player.

Resonance occurs widely in nature, and is exploited in many manmade devices. It is the mechanism by which virtually all sinusoidal waves and vibrations are generated. Many sounds we hear, such as when hard objects of metal, glass, or wood are struck, are caused by brief resonant vibrations in the object. Light and other short wavelength electromagnetic radiation is produced by resonance on an atomic scale, such as electrons in atoms.

Examples

Swing

Pushing a person in a swing is a common example of resonance. The loaded swing, a pendulum, has a natural frequency of oscillation, its resonant frequency, and resists being pushed at a faster or slower rate.

A familiar example is a playground swing, which acts as a pendulum. Pushing a person in a swing in time with the natural interval of the swing (its resonant frequency) makes the swing go higher and higher (maximum amplitude), while attempts to push the swing at a faster or slower tempo produce smaller arcs. This is because the energy the swing absorbs is maximized when the pushes match the swing's natural oscillations.

Mechanical and Acoustic Resonance

Mechanical resonance is the tendency of a mechanical system to absorb more energy when the frequency of its oscillations matches the system's natural frequency of vibration than it does at other frequencies. It may cause violent swaying motions and even catastrophic failure in improperly constructed structures including bridges, buildings, trains, and aircraft. When designing objects, engineers must ensure the mechanical resonance frequencies of the component parts do not match driving vibrational frequencies of motors or other oscillating parts, a phenomenon known as resonance disaster.

Avoiding resonance disasters is a major concern in every building, tower, and bridge construction project. As a countermeasure, shock mounts can be installed to absorb resonant frequencies and thus dissipate the absorbed energy. The Taipei 101 building relies on a 660-tonne pendulum (730-short-ton)—a tuned mass damper—to cancel resonance. Furthermore, the structure is designed to resonate at a frequency that does not typically occur. Buildings in seismic zones are often constructed to take into account the oscillating frequencies of expected ground motion. In addition, engineers designing objects having engines must ensure that the mechanical resonant frequencies of the component parts do not match driving vibrational frequencies of the motors or other strongly oscillating parts.

Clocks keep time by mechanical resonance in a balance wheel, pendulum, or quartz crystal.

The cadence of runners has been hypothesized to be energetically favorable due to resonance between the elastic energy stored in the lower limb and the mass of the runner.

Acoustic resonance is a branch of mechanical resonance that is concerned with the mechanical vibrations across the frequency range of human hearing, in other words sound. For humans, hearing is normally limited to frequencies between about 20 Hz and 20,000 Hz (20 kHz),

Acoustic resonance is an important consideration for instrument builders, as most acoustic instruments use resonators, such as the strings and body of a violin, the length of tube in a flute, and the shape of, and tension on, a drum membrane.

Like mechanical resonance, acoustic resonance can result in catastrophic failure of the object at resonance. The classic example of this is breaking a wine glass with sound at the precise resonant frequency of the glass, although this is difficult in practice.

Tacoma Narrows Bridge

The dramatically visible, rhythmic twisting that resulted in the 1940 collapse of "Galloping Gertie", the original Tacoma Narrows Bridge, is misleadingly characterized as an example of resonance phenomenon in certain textbooks. The catastrophic vibrations that destroyed the bridge were not due to simple mechanical resonance, but to a more complicated interaction between the bridge and the winds passing through it—a phenomenon known as aeroelastic flutter, which is a kind of "self-sustaining vibration" as referred to in the nonlinear theory of vibrations. Robert H. Scanlan, father of bridge aerodynamics, has written an article about this misunderstanding.

Electrical Resonance

Electrical resonance occurs in an electric circuit at a particular *resonant frequency* when the impedance of the circuit is at a minimum in a series circuit or at maximum in a parallel circuit (or when the transfer function is at a maximum).

Optical Resonance

An optical cavity, also called an *optical resonator*, is an arrangement of mirrors that forms a standing wave cavity resonator for light waves. Optical cavities are a major component of lasers, surrounding the gain medium and providing feedback of the laser light. They are also used in optical parametric oscillators and some interferometers. Light confined in the cavity reflects multiple times producing standing waves for certain resonant frequencies. The standing wave patterns produced are called "modes". Longitudinal modes differ only in frequency while transverse modes differ for different frequencies and have different intensity patterns across the cross-section of the beam. Ring resonators and whispering galleries are examples of optical resonators that do not form standing waves.

Different resonator types are distinguished by the focal lengths of the two mirrors and the distance between them; flat mirrors are not often used because of the difficulty of aligning them precisely. The geometry (resonator type) must be chosen so the beam remains stable, i.e., the beam size does not continue to grow with each reflection. Resonator types are also designed to meet other criteria such as minimum beam waist or having no focal point (and therefore intense light at that point) inside the cavity.

Optical cavities are designed to have a very large Q factor. A beam reflects a large number of times with little attenuation—therefore the frequency line width of the beam is small compared to the frequency of the laser.

Additional optical resonances are guided-mode resonances and surface plasmon resonance, which result in anomalous reflection and high evanescent fields at resonance. In this case, the resonant modes are guided modes of a waveguide or surface plasmon modes of a dielectric-metallic interface. These modes are usually excited by a subwavelength grating.

Orbital Resonance

In celestial mechanics, an orbital resonance occurs when two orbiting bodies exert a regular, periodic gravitational influence on each other, usually due to their orbital periods being related by a ratio of two small integers. Orbital resonances greatly enhance the mutual gravitational influence of the bodies. In most cases, this results in an *unstable* interaction, in which the bodies exchange momentum and shift orbits until the resonance no longer exists. Under some circumstances, a resonant system can be stable and self-correcting, so that the bodies remain in resonance. Examples are the 1:2:4 resonance of Jupiter's moons Ganymede, Europa, and Io, and the 2:3 resonance between Pluto and Neptune. Unstable resonances with Saturn's inner moons give rise to gaps in the rings of Saturn. The special case of 1:1 resonance (between bodies with similar orbital radii) causes large Solar System bodies to clear the neighborhood around their orbits by ejecting nearly everything else around them; this effect is used in the current definition of a planet.

Atomic, Particle, and Molecular Resonance

Nuclear magnetic resonance (NMR) is the name given to a physical resonance phenomenon involving the observation of specific quantum mechanical magnetic properties of an atomic nucleus in the presence of an applied, external magnetic field. Many scientific techniques exploit NMR phenomena to study molecular physics, crystals, and non-crystalline materials through NMR spectroscopy. NMR is also routinely used in advanced medical imaging techniques, such as in magnetic resonance imaging (MRI).

NMR Magnet at HWB-NMR, Birmingham, UK. In its strong 21.2-tesla field, the proton resonance is at 900 MHz.

All nuclei containing odd numbers of nucleons have an intrinsic magnetic moment and angular momentum. A key feature of NMR is that the resonant frequency of a particular substance is directly proportional to the strength of the applied magnetic field. It is this feature that is exploited in imaging techniques; if a sample is placed in a non-uniform magnetic field then the resonant frequencies of the sample's nuclei depend on where in the field they are located. Therefore, the particle can be located quite precisely by its resonant frequency.

Electron paramagnetic resonance, otherwise known as *Electron Spin Resonance* (ESR) is a spectroscopic technique similar to NMR, but uses unpaired electrons instead. Materials for which this can be applied are much more limited since the material needs to both have an unpaired spin and be paramagnetic.

The Mössbauer effect is the resonant and recoil-free emission and absorption of gamma ray photons by atoms bound in a solid form.

Resonance in particle physics appears in similar circumstances to classical physics at the level of quantum mechanics and quantum field theory. However, they can also be thought of as unstable particles, with the formula above valid if Γ is the decay rate and Ω replaced by the particle's mass M. In that case, the formula comes from the particle's propagator, with its mass replaced by the complex number $M + i\Gamma$. The formula is further related to the particle's decay rate by the optical theorem.

International Space Station

The rocket engines for the International Space Station (ISS) are controlled by an autopilot. Ordinarily, uploaded parameters for controlling the engine control system for the Zvezda module make the rocket engines boost the International Space Station to a higher orbit. The rocket engines are hinge-mounted, and ordinarily the crew doesn't notice the operation. On January 14, 2009, however, the uploaded parameters made the autopilot swing the rocket engines in larger and larger oscillations, at a frequency of 0.5 Hz. These oscillations were captured on video, and lasted for 142 seconds.

Other Examples

- Timekeeping mechanisms of modern clocks and watches, e.g., the balance wheel in a mechanical watch and the quartz crystal in a quartz watch
- Tidal resonance of the Bay of Fundy
- Acoustic resonances of musical instruments and the human vocal tract
- Shattering of a crystal wineglass when exposed to a musical tone of the right pitch (its resonant frequency)
- Friction idiophones, such as making a glass object (glass, bottle, vase) vibrate by rubbing around its rim with a fingertip
- Electrical resonance of tuned circuits in radios and TVs that allow radio frequencies to be selectively received
- Creation of coherent light by optical resonance in a laser cavity
- Orbital resonance as exemplified by some moons of the solar system's gas giants
- Material resonances in atomic scale are the basis of several spectroscopic techniques that are used in condensed matter physics
    - Electron spin resonance
    - Mössbauer effect
    - Nuclear magnetic resonance

Theory

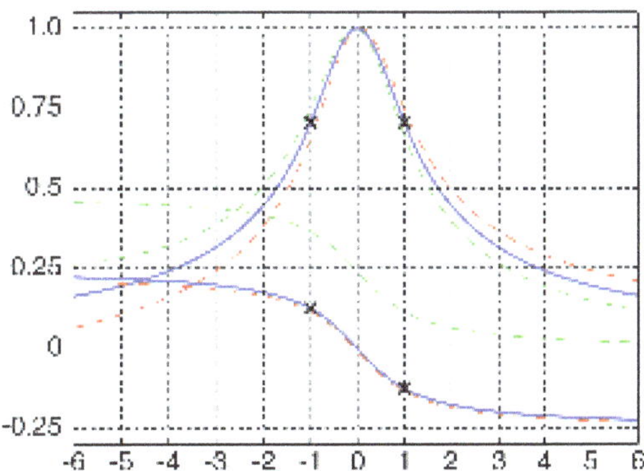

"Universal Resonance Curve", a symmetric approximation to the normalized response of a resonant circuit; abscissa values are deviation from center frequency, in units of center frequency divided by 2Q; ordinate is relative amplitude, and phase in cycles; dashed curves compare the range of responses of real two-pole circuits for a *Q* value of 5; for higher *Q* values, there is less deviation from the universal curve. Crosses mark the edges of the 3 dB bandwidth (gain 0.707, phase shift 45° or 0.125 cycle).

The exact response of a resonance, especially for frequencies far from the resonant frequency, depends on the details of the physical system, and is usually not exactly symmetric about the resonant frequency, as illustrated for the simple harmonic oscillator above. For a lightly damped linear oscillator with a resonance frequency Ω, the *intensity* of oscillations I when the system is driven with a driving frequency ω is typically approximated by a formula that is symmetric about the resonance frequency:

$$I(\omega) \propto \frac{\left(\frac{\Gamma}{2}\right)^2}{(\omega-\Omega)^2+\left(\frac{\Gamma}{2}\right)^2}.$$

The intensity is defined as the square of the amplitude of the oscillations. This is a Lorentzian function, or Cauchy distribution, and this response is found in many physical situations involving resonant systems. Γ is a parameter dependent on the damping of the oscillator, and is known as the *linewidth* of the resonance. Heavily damped oscillators tend to have broad linewidths, and respond to a wider range of driving frequencies around the resonant frequency. The linewidth is inversely proportional to the Q factor, which is a measure of the sharpness of the resonance.

In electrical engineering, this approximate symmetric response is known as the *universal resonance curve*, a concept introduced by Frederick E. Terman in 1932 to simplify the approximate analysis of radio circuits with a range of center frequencies and Q values.

Resonators

A physical system can have as many resonant frequencies as it has degrees of freedom; each degree of freedom can vibrate as a harmonic oscillator. Systems with one degree of freedom, such as a mass on a spring, pendulums, balance wheels, and LC tuned circuits have one resonant frequency. Systems with two degrees of freedom, such as coupled pendulums and resonant transformers can have two resonant frequencies. As the number of coupled harmonic oscillators grows, the time it takes to transfer energy from one to the next becomes significant. The vibrations in them begin to travel through the coupled harmonic oscillators in waves, from one oscillator to the next.

Extended objects that can experience resonance due to vibrations inside them are called resonators, such as organ pipes, vibrating strings, quartz crystals, microwave and laser cavities. Since these can be viewed as being made of millions of coupled moving parts (such as atoms), they can have millions of resonant frequencies. The vibrations inside them travel as waves, at an approximately constant velocity, bouncing back and forth between the sides of the resonator. If the distance between the sides is d, the length of a roundtrip is $2d$. To cause resonance, the phase of a sinusoidal wave after a roundtrip must be equal to the initial phase, so the waves reinforce the oscillation. So the condition for resonance in a resonator is that the roundtrip distance, $2d$, be equal to an integer number of wavelengths λ of the wave:

$$2d = N\lambda, \qquad N \in \{1,2,3,\ldots\}$$

If the velocity of a wave is v, the frequency is $f = \frac{v}{\lambda}$ so the resonant frequencies are:

$$f = \frac{Nv}{2d} \qquad N \in \{1,2,3,\ldots\}$$

So the resonant frequencies of resonators, called normal modes, are equally spaced multiples of a lowest frequency called the fundamental frequency. The multiples are often called overtones. There may be several such series of resonant frequencies, corresponding to different modes of oscillation.

Q Factor

The *Q* factor or *quality factor* is a dimensionless parameter that describes how under-damped an oscillator or resonator is, or equivalently, characterizes a resonator's bandwidth relative to its center frequency. Higher *Q* indicates a lower rate of energy loss relative to the stored energy of the oscillator, i.e., the oscillations die out more slowly. A pendulum suspended from a high-quality bearing, oscillating in air, has a high *Q*, while a pendulum immersed in oil has a low *Q*. To sustain a system in resonance in constant amplitude by providing power externally, the energy provided in each cycle must be less than the energy stored in the system (i.e., the sum of the potential and kinetic) by a factor of $\frac{Q}{2\pi}$. Oscillators with high-quality factors have low damping, which tends to make them ring longer.

Sinusoidally driven resonators having higher Q factors resonate with greater amplitudes (at the resonant frequency) but have a smaller range of frequencies around the frequency at which they resonate. The range of frequencies at which the oscillator resonates is called the bandwidth. Thus, a high-*Q* tuned circuit in a radio receiver would be more difficult to tune, but would have greater selectivity, it would do a better job of filtering out signals from other stations that lie nearby on the spectrum. High Q oscillators operate over a smaller range of frequencies and are more stable.

The quality factor of oscillators varies substantially from system to system. Systems for which damping is important (such as dampers keeping a door from slamming shut) have Q= $\frac{1}{2}$. Clocks, lasers, and other systems that need either strong resonance or high frequency stability need high-quality factors. Tuning forks have quality factors around $Q = 1000$. The quality factor of atomic clocks and some high-*Q* lasers can reach as high as 10^{11} and higher.

There are many alternate quantities used by physicists and engineers to describe how damped an oscillator is that are closely related to its quality factor. Important examples include: the damping ratio, relative bandwidth, linewidth, and bandwidth measured in octaves.

Electrical Circuits

Electrical Circuits are the most common technological application where we see resonances. The LCR circuit shown in Figure characterizes the typical situation. The circuits includes a signal generator which produces an AC signal of voltage amplitude V at frequency ω. Applying Kirchoff's Law to this circuit we have,

$$-Ve^{i\omega t} + L\frac{dI}{dt} + \frac{q}{C} + RI = 0,$$

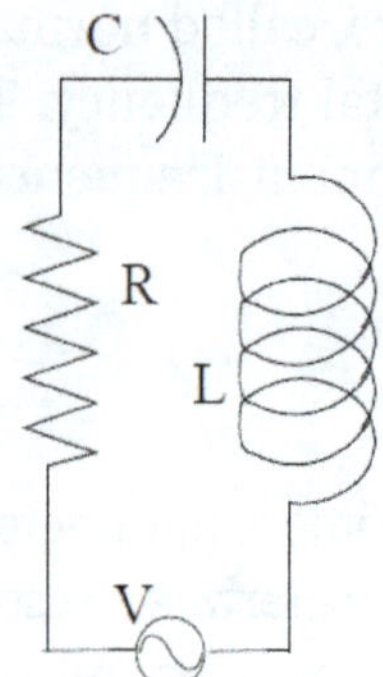

which may be written solely in terms of the charge q on the capacitor as with $I = \frac{dq}{dt}$

$$L\ddot{q} + R\dot{q} + \frac{q}{C} = Ve^{i\omega t},.$$

We see that this is a damped oscillator with an external sinusoidal force. The equation governing this is

$$\ddot{\tilde{q}} + 2\beta\dot{\tilde{q}} + \omega_0^2\tilde{q} = \tilde{\upsilon}e^{i\omega t}$$

Where $\omega_0^2 = 1/LC, \beta = R/2L \; and \; \upsilon = (V/L).$

We next consider the power dissipated in this circuit. The resistance is the only circuit element which draws power. We proceed to calculate this by calculating the impedence

$$\tilde{Z}(\omega) = i\omega L - \frac{i}{\omega C} + R$$

which varies with frequency. The voltage and current are related as $\tilde{\text{V}}=\tilde{\text{I}}\tilde{\text{Z}}$, which gives the current

$$\tilde{\text{I}}=\frac{\tilde{\text{V}}}{i(\omega L - 1/\omega C) + R}.$$

The average power dissipated may be calculated as $\langle \text{P}(\omega) \rangle = \text{R}\tilde{\text{I}}\tilde{\text{I}}^*/2$ which is

$$\langle \text{P}(\omega) \rangle = \frac{\omega^2}{(\omega_0^2 - \omega^2)^2 + 4\beta^2\omega^2}\left(\frac{RV^2}{2L^2}\right)$$

a Lorentzian as expected.

The Raman Effect

Light of frequency ν is incident on a target. If the emergent light is analysed through a spectrometer it is found that theere are components at two new frequencies $\nu - \Delta\nu$ and $\nu + \Delta\nu$ known as the Stokes and anti-Stokes lines respectively. This phenomenon was discovered by Sir C.V. Raman and it is known as the Raman Effect.

As an example, we consider light of frequency $\nu = 6.0 \times 10^{14}\, Hz$ incident on benzene which is a liquid. It is found that there are three different pairs of Stokes and anti-Stokes lines in the spectrum. It is possible to associate each of these new pair of lines with different oscillations of the benzene molecule. The vibrations of a complex system like benzene can be decomposed into different normal modes, each of which behaves like a simple harmonic oscillator with its own natural frequency. There is a separate Raman line associated with eah of these different modes. A closer look at these spectral lines shows them to have a finite width, the shape being a Lorentzian corresponding to the resonance of a damped harmonic oscillators. Figure shows the Raman line corresponding to the bending mode of benzene.

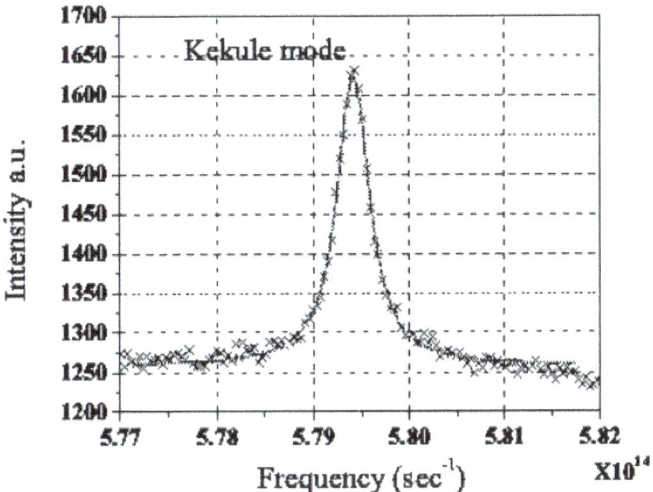

Coupled Oscillators

Consider two idential simple harmonic oscillators of mass m and spring constant k as shown in Figure (a.). The two oscillators are independent with

$$x_0(t) = a_0 \cos(\omega t + \phi_0)$$

And

$$x_1(t) = a_1 \cos(\omega t + \phi_1)$$

where they both oscillate with the same frequency $\omega = \sqrt{\frac{k}{m}}$. The amplitudes a_0, a_1 and the phases ϕ_0, ϕ_1 of the two oscillators are in no way interdependent. The question which we take up for discussion here is what happens if the two masses are coupled by a third spring as shown in Figure (b.).

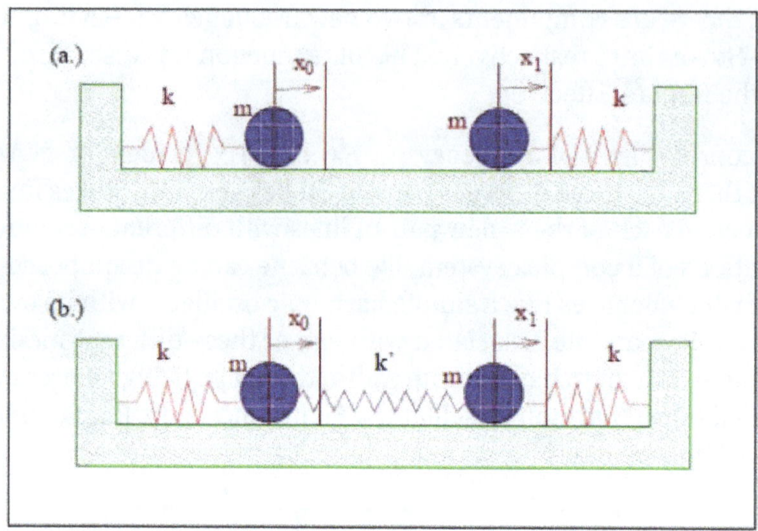

This shows two identical spring-mass systems.
In (a.) the two oscillators are independent whereas in (b.) they are coupled through an extra spring.

The motion of the two oscillators is now coupled through the third spring of spring constant k'. It is clear that the oscillation of one oscillator affects the second. The phases and amplitudes of the two oscillators are no longer independent and the frequency of oscillation is also modified. We proceed to calculate these effects below.

The equations governing the coupled oscillators are

$$m\frac{d^2x_0}{dt^2} = -kx_0 - k'(x_0 - x_1)$$

And

$$m\frac{d^2x_1}{dt^2} = -kx_1 - k'(x_1 - x_0)$$

Normal Modes

The technique to solve such *coupled differential equations* is to identify linear combinations of x_0 and x_1 for which the equations become decoupled. In this case it is very easy to identify such variables

$$q_0 = \frac{x_0 + x_1}{2} \text{ and } q_1 = \frac{x_0 - x_1}{2}$$

These are referred to as as the *normal modes* (or *eigen modes*) of the system and the equations governing them are

$$m\frac{d^2q_0}{dt^2} = -kq_0$$

and

$$m\frac{d^2q_1}{dt^2} = -(k+2k')q_1.$$

The two normal modes execute simple harmonic oscillations with respective angular frequencies

$$\omega_0 = \sqrt{\frac{k}{m}} \; and \; \omega_1 = \sqrt{\frac{k+2k'}{m}}$$

In this case the normal modes lend themselves to a simple physical interpretation.

The normal mode q_0 represents the center of mass. The center of mass behaves as if it were a particle of mass $2m$ attached to two springs and its oscillation frequency is the same as that of the individual decoupled oscillators $\omega_0 = \sqrt{\frac{2k}{2m}}$.

The normal mode q_1 represents the relative motion of he two masses which leaves the center of mass unchanged. This can be thought of as the motion of two particles of mass m connected to a spring of spring constant $\tilde{k} = (k+2k')/2$ as shown in Figure. The oscillation frequency of this normal mode $\omega_1 = \sqrt{\frac{k+2k'}{m}}$ is always higher than that of the individual uncoupled oscillators (or the center of mass). The modes q_0 and q_1 are often referred to as the slow mode and the fast mode respectively.

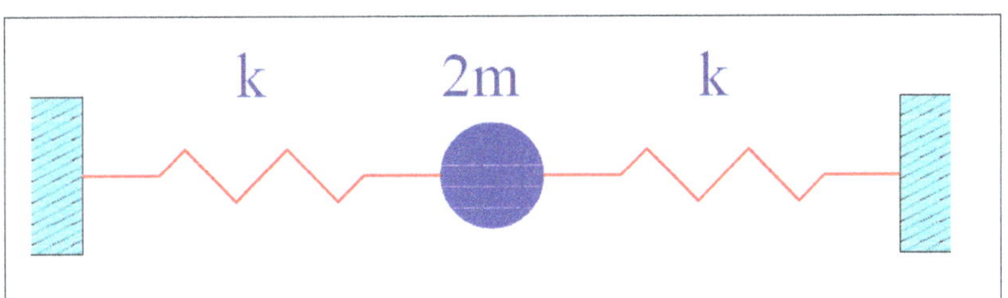

This shows the spring mass equivalent of the normal mode q_0 which corresponds to the center of mass.

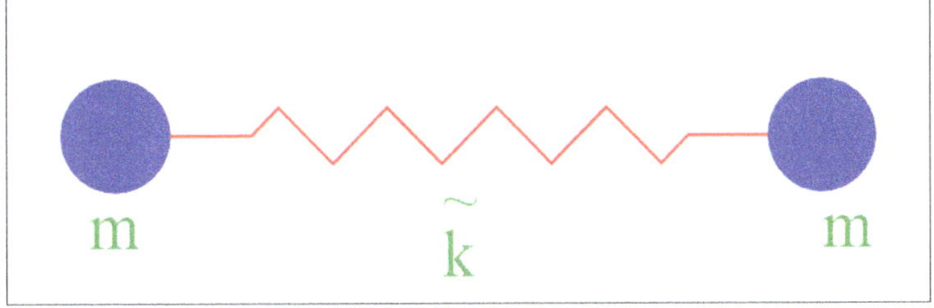

This shows the spring mass equivalent of the normal mode q1 which corresponds to two particles connected through a spring.

We may interpret q_0 as a mode of oscillation where the two masses oscillate with exactly the same phase, and q_0 as a mode where they have a phase difference of π. Recollect that the phases of the two masses are independent when the two masses are not coupled. Introducing a coupling causes the phases to be interdependent.

The normal modes have solutions

$$\tilde{q}_0(t) = \tilde{A}_0 \ e^{i\omega_0 t}$$

$$\tilde{q}_1(t) = \tilde{A}_1 \ e^{i\omega_1 t}$$

where it should be bourne in mind that $\tilde{A}_0$ and $\tilde{A}_1$ are complex numbers with both amplitude and phase ie. $\tilde{A}_0 = A_0 e^{i\phi_0}$ etc. We then have the solutions

$$\tilde{x}_0(t) = \tilde{A}_0 \ e^{i\omega_0 t} + \tilde{A}_1 \ e^{i\omega_1 t}$$

$$\tilde{x}_1(t) = \tilde{A}_0 \ e^{i\omega_0 t} - \tilde{A}_1 \ e^{i\omega_1 t}$$

The complex amplitudes $\tilde{A}_1$ and $\tilde{A}_2$ have to be determined from the initial conditions, four initial conditions are required in total.

Resonance

As an example we consider a situation where the two particles are initially at rest in the equilibrium position. The particle x_0 is given a small displacement

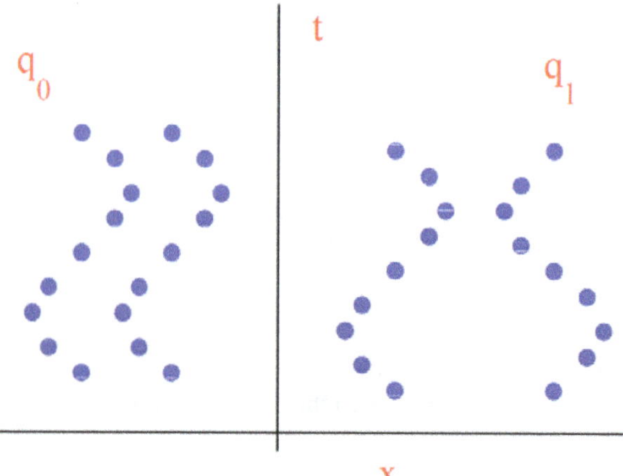

This shows the motion corresponding to the two normal modes q0 and q1 respectively.

a_0 and then left to oscillate. Using this to determine $\tilde{A}_1$ and $\tilde{A}_2$, we finally have

$$x_0(t) = \frac{a_0}{2}\left[\cos \omega_0 t + \cos \omega_1 t\right]$$

And

$$x_1(t) = \frac{a_0}{2}\left[\cos \omega_0 t - \cos \omega_1 t\right]$$

The solution can also be written as

$$x_0(t) = a_0 \cos\left[\left(\frac{\omega_1 - \omega_0}{2}\right)t\right]\cos\left[\left(\frac{\omega_0 + \omega_1}{2}\right)t\right]$$

$$x_1(t) = a_0 \sin\left[\left(\frac{\omega_1 - \omega_0}{2}t\right)\sin\left(\frac{\omega_0 + \omega_1}{2}\right)t\right]$$

It is interesting to consider $k' \ll k$ where the two oscillators are weakly coupled. In this limit

$$\omega_1 = \sqrt{\frac{k}{m}\left(1 + \frac{2k'}{k}\right)} \approx \omega_0 + \frac{k'}{k}\omega_0$$

and we have solutions

$$x_0(t) = \left[a_0 \cos\left(\frac{k'}{2k}\omega_0 t\right)\right]\cos \omega_0 t$$

And

$$x_1(t) = \left[a_0 \sin\left(\frac{k'}{2k}\omega_0 t\right)\right]\sin \omega_0 t.$$

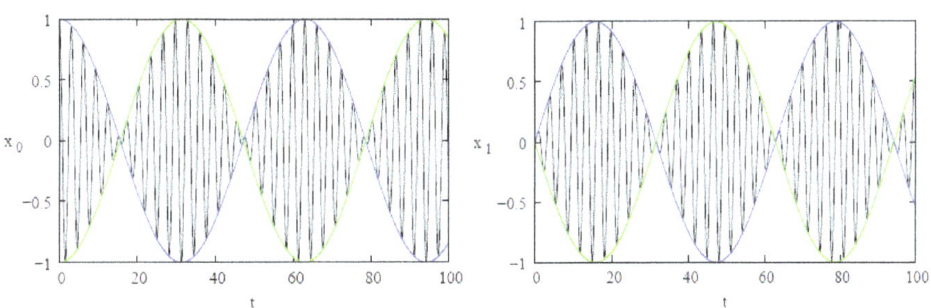

This shows the motion of x_0 and x_1.

The solution is shown in Figure. We can think of the motion as an oscillation with ω_0 where the amplitude undergoes a slow modulation at angular frequency $\frac{k'}{2k}\omega_0$. The oscillations of the two particles are out of phase and are slowly transferred from the particle which receives the initial displacement to the particle originally at rest, and then back again.

Wave

In physics, a wave is an oscillation accompanied by a transfer of energy that travels through a medium (space or mass). Frequency refers to the addition of time. *Wave motion* transfers energy from one point to another, which displace particles of the transmission medium–that is, with little or no associated mass transport. Waves consist, instead, of oscillations or vibrations (of a physical quantity), around almost fixed locations.

There are two main types of waves. Mechanical waves propagate through a medium, and the substance of this medium is deformed. Restoring forces then reverse the deformation. For example, sound waves propagate via air molecules colliding with their neighbors. When the molecules collide, they also bounce away from each other (a restoring force). This keeps the molecules from continuing to travel in the direction of the wave.

The second main type, electromagnetic waves, do not require a medium. Instead, they consist of periodic oscillations of electrical and magnetic fields originally generated by charged particles, and can therefore travel through a vacuum. These types vary in wavelength, and include radio waves, microwaves, infrared radiation, visible light, ultraviolet radiation, X-rays and gamma rays.

Waves are described by a wave equation which sets out how the disturbance proceeds over time. The mathematical form of this equation varies depending on the type of wave. Further, the behavior of particles in quantum mechanics are described by waves. In addition, gravitational waves also travel through space, which are a result of a vibration or movement in gravitational fields.

A wave can be transverse, where a disturbance creates oscillations that are perpendicular to the propagation of energy transfer, or longitudinal: the oscillations are parallel to the direction of energy propagation. While mechanical waves can be both transverse and longitudinal, all electromagnetic waves are transverse in free space.

General Features

Surface waves in water showing water ripples

A single, all-encompassing definition for the term *wave* is not straightforward. A vibration can be defined as a *back-and-forth* motion around a reference value. However, a vibration is not necessarily a wave. An attempt to define the necessary and sufficient characteristics that qualify a phenomenon as a *wave* results in a blurred line.

The term *wave* is often intuitively understood as referring to a transport of spatial disturbances that are generally not accompanied by a motion of the medium occupying this space as a whole. In a wave, the energy of a vibration is moving away from the source in the form of a disturbance within the surrounding medium (Hall 1980, p. 8). However, this motion is problematic for a standing wave (for example, a wave on a string), where energy is moving in both directions equally, or for electromagnetic (e.g., light) waves in a vacuum, where the concept of medium does not apply and interaction with a target is the key to wave detection and practical applications. There are water waves on the ocean surface; gamma waves and light waves emitted by the Sun; microwaves used in microwave ovens and in radar equipment; radio waves broadcast by radio stations; and sound waves generated by radio receivers, telephone handsets and living creatures (as voices), to mention only a few wave phenomena.

It may appear that the description of waves is closely related to their physical origin for each specific instance of a wave process. For example, acoustics is distinguished from optics in that sound waves are related to a mechanical rather than an electromagnetic wave transfer caused by vibration. Concepts such as mass, momentum, inertia, or elasticity, become therefore crucial in describing acoustic (as distinct from optic) wave processes. This difference in origin introduces certain wave characteristics particular to the properties of the medium involved. For example, in the case of air: vortices, radiation pressure, shock waves etc.; in the case of solids: Rayleigh waves, dispersion; and so on....

Other properties, however, although usually described in terms of origin, may be generalized to all waves. For such reasons, wave theory represents a particular branch of physics that is concerned with the properties of wave processes independently of their physical origin. For example, based on the mechanical origin of acoustic waves, a moving disturbance in space–time can exist if and only if the medium involved is neither infinitely stiff nor infinitely pliable. If all the parts making up a medium were rigidly *bound*, then they would all vibrate as one, with no delay in the transmission of the vibration and therefore no wave motion. On the other hand, if all the parts were independent, then there would not be any transmission of the vibration and again, no wave motion. Although the above statements are meaningless in the case of waves that do not require a medium, they reveal a characteristic that is relevant to all waves regardless of origin: within a wave, the phase of a vibration (that is, its position within the vibration cycle) is different for adjacent points in space because the vibration reaches these points at different times.

Mathematical Description of One-dimensional Waves

Wave Equation

Consider a traveling transverse wave (which may be a pulse) on a string (the medium). Consider the string to have a single spatial dimension. Consider this wave as traveling

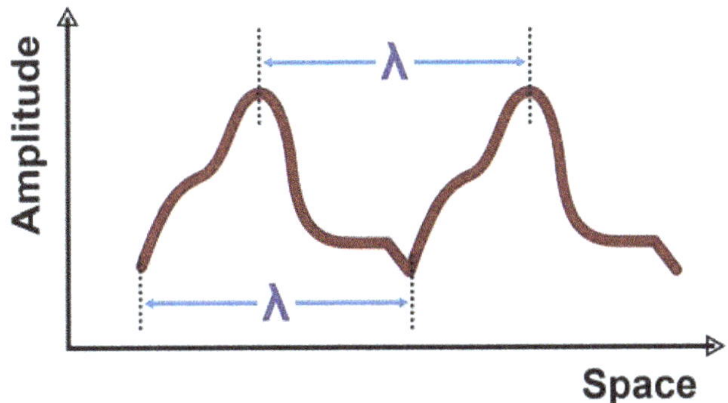

Wavelength λ, can be measured between any two corresponding points on a waveform

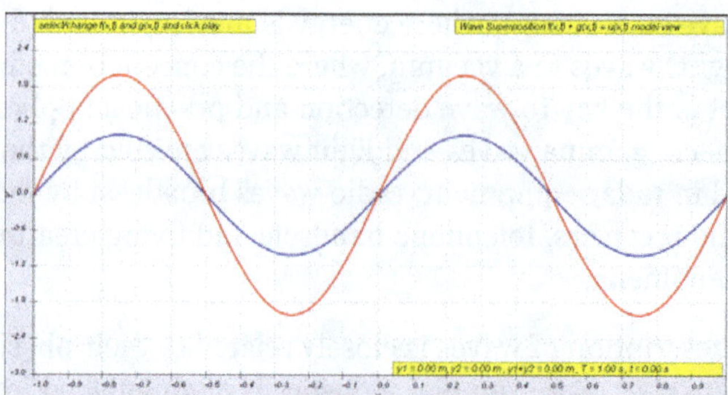

Animation of two waves, the green wave moves to the right while blue wave moves to the left, the net red wave amplitude at each point is the sum of the amplitudes of the individual waves. Note that f(x,t) + g(x,t) = u(x,t)

- in the x direction in space. E.g., let the positive x direction be to the right, and the negative x direction be to the left.
- with constant amplitude u
- with constant velocity v, where v is
    - o independent of wavelength (no dispersion)
    - o independent of amplitude (linear media, not nonlinear).
- with constant waveform, or shape

This wave can then be described by the two-dimensional functions

$u(x,t) = F(x - v\,t)$ (waveform F traveling to the right)

$u(x,t) = G(x + v\,t)$ (waveform G traveling to the left)

or, more generally, by d'Alembert's formula:

$$u(x,t) = F(x - vt) + G(x + vt).$$

representing two component waveforms F and G traveling through the medium in opposite directions. A generalized representation of this wave can be obtained as the partial differential equation

$$\frac{1}{v^2}\frac{\partial^2 u}{\partial t^2} = \frac{\partial^2 u}{\partial x^2}.$$

General solutions are based upon Duhamel's principle.

Wave Forms

The form or shape of F in d'Alembert's formula involves the argument $x - vt$. Constant values of this argument correspond to constant values of F, and these constant values occur if x increases at the same rate that vt increases. That is, the wave shaped like the function F will move in the positive x-direction at velocity v (and G will propagate at the same speed in the negative x-direction).

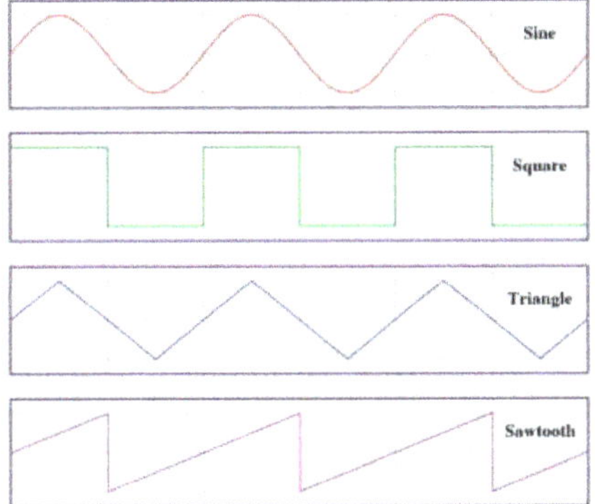

Sine, square, triangle and sawtooth waveforms.

In the case of a periodic function F with period λ, that is, $F(x + \lambda - vt) = F(x - vt)$, the periodicity of F in space means that a snapshot of the wave at a given time t finds the wave varying periodically in space with period λ (the wavelength of the wave). In a similar fashion, this periodicity of F implies a periodicity in time as well: $F(x - v(t + T)) = F(x - vt)$ provided $vT = \lambda$, so an observation of the wave at a fixed location x finds the wave undulating periodically in time with period $T = \lambda/v$.

Amplitude and Modulation

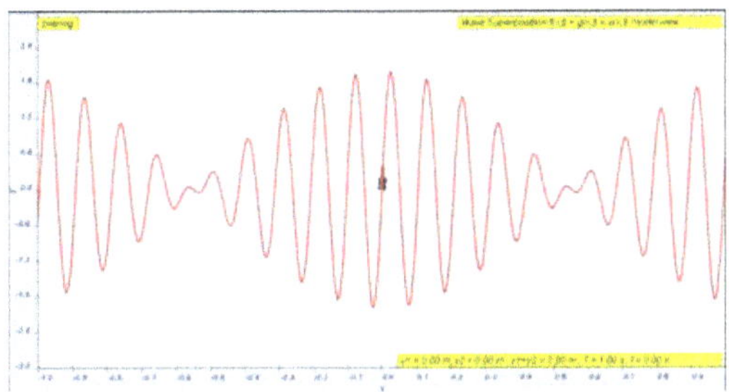

Amplitude modulation can be achieved through f(x,t) = 1.00*sin(2*pi/0.10*(x-1.00*t)) and g(x,t) = 1.00*sin(2*pi/0.11*(x-1.00*t))only the resultant is visible to improve clarity of waveform.

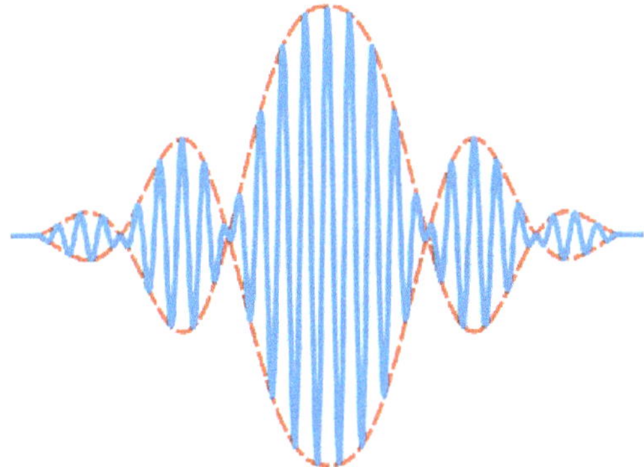

Illustration of the *envelope* (the slowly varying red curve) of an amplitude-modulated wave. The fast varying blue curve is the *carrier* wave, which is being modulated.

The amplitude of a wave may be constant (in which case the wave is a *c.w.* or *continuous wave*), or may be *modulated* so as to vary with time and/or position. The outline of the variation in amplitude is called the *envelope* of the wave. Mathematically, the modulated wave can be written in the form:

$$u(x,t) = A(x,t)\sin(kx - \omega t + \phi),$$

where $A(x, t)$ is the amplitude envelope of the wave, k is the *wavenumber* and ϕ is the *phase*. If the group velocity v_g is wavelength-independent, this equation can be simplified as:

$$u(x,t) = A(x - v_g\ t)\sin(kx - \omega t + \phi),$$

showing that the envelope moves with the group velocity and retains its shape. Otherwise, in cases where the group velocity varies with wavelength, the pulse shape changes in a manner often described using an *envelope equation.*

Phase Velocity and Group Velocity

Frequency dispersion in groups of gravity waves on the surface of deep water. The red dot moves with the phase velocity, and the green dots propagate with the group velocity.

There are two velocities that are associated with waves, the phase velocity and the group velocity. To understand them, one must consider several types of waveform. For simplification, examination is restricted to one dimension.

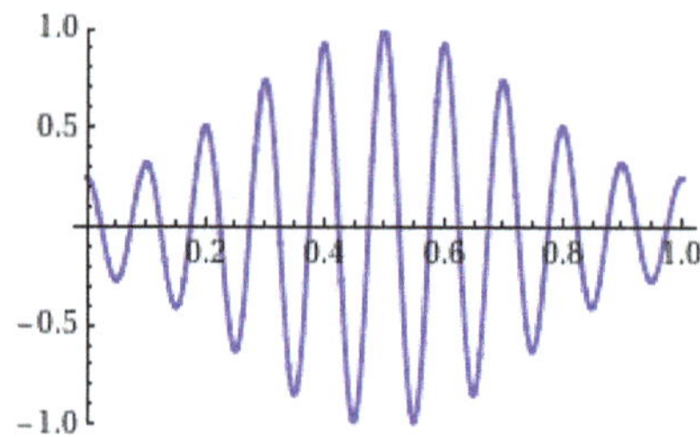

This shows a wave with the Group velocity and Phase velocity going in different directions.

The most basic wave (a form of plane wave) may be expressed in the form:

$$\psi(x,t) = Ae^{i(kx - \omega t)},$$

which can be related to the usual sine and cosine forms using Euler's formula. Rewriting the argument, $kx - \omega t = \left(\frac{2\pi}{\lambda}\right)(x - vt)$, makes clear that this expression describes a vibration of wavelength $\lambda = \frac{2\pi}{k}$ traveling in the x-direction with a constant *phase velocity* $v_p = \frac{\omega}{k}$.

The other type of wave to be considered is one with localized structure described by an envelope, which may be expressed mathematically as, for example:

$$\psi(x,t) = \int_{-\infty}^{\infty} dk_1\ A(k_1)\ e^{i(k_1 x - \omega t)},$$

where now $A(k_1)$ (the integral is the inverse Fourier transform of A(k1)) is a function exhibiting a sharp peak in a region of wave vectors Δk surrounding the point $k_1 = k$. In exponential form:

$$A = A_o(k_1)e^{i\alpha(k_1)} ,$$

with A_o the magnitude of A. For example, a common choice for A_o is a Gaussian wave packet:

$$A_o(k_1) = N\,e^{-\sigma^2(k_1-k)^2/2} ,$$

where σ determines the spread of k_1-values about k, and N is the amplitude of the wave.

The exponential function inside the integral for ψ oscillates rapidly with its argument, say $\varphi(k_1)$, and where it varies rapidly, the exponentials cancel each other out, interfere destructively, contributing little to ψ. However, an exception occurs at the location where the argument φ of the exponential varies slowly. (This observation is the basis for the method of stationary phase for evaluation of such integrals.) The condition for φ to vary slowly is that its rate of change with k_1 be small; this rate of variation is:

$$\left.\frac{d\varphi}{dk_1}\right|_{k_1=k} = x - t\left.\frac{d\omega}{dk_1}\right|_{k_1=k} + \left.\frac{d\alpha}{dk_1}\right|_{k_1=k} ,$$

where the evaluation is made at $k_1 = k$ because $A(k_1)$ is centered there. This result shows that the position x where the phase changes slowly, the position where ψ is appreciable, moves with time at a speed called the *group velocity*:

$$v_g = \frac{d\omega}{dk} .$$

The group velocity therefore depends upon the dispersion relation connecting ω and k. For example, in quantum mechanics the energy of a particle represented as a wave packet is $E = \hbar\omega = (\hbar k)^2/(2m)$. Consequently, for that wave situation, the group velocity is

$$v_g = \frac{\hbar k}{m} ,$$

showing that the velocity of a localized particle in quantum mechanics is its group velocity. Because the group velocity varies with k, the shape of the wave packet broadens with time, and the particle becomes less localized. In other words, the velocity of the constituent waves of the wave packet travel at a rate that varies with their wavelength, so some move faster than others, and they cannot maintain the same interference pattern as the wave propagates.

Sinusoidal Waves

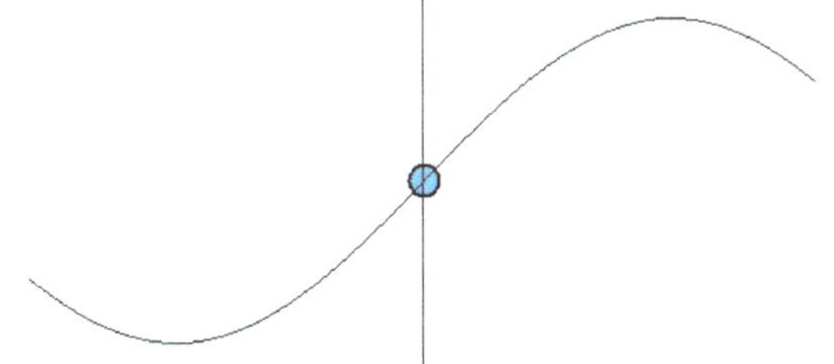

Sinusoidal waves correspond to simple harmonic motion.

Mathematically, the most basic wave is the (spatially) one-dimensional sine wave (or *harmonic wave* or *sinusoid*) with an amplitude u described by the equation:

$$u(x,t) = A\sin(kx - \omega t + \phi),$$

where

- A is the maximum amplitude of the wave, maximum distance from the highest point of the disturbance in the medium (the crest) to the equilibrium point during one wave cycle. In the illustration to the right, this is the maximum vertical distance between the baseline and the wave.
- x is the space coordinate
- t is the time coordinate
- k is the wavenumber
- ω is the angular frequency
- ϕ is the phase constant.

The units of the amplitude depend on the type of wave. Transverse mechanical waves (e.g., a wave on a string) have an amplitude expressed as a distance (e.g., meters), longitudinal mechanical waves (e.g., sound waves) use units of pressure (e.g., pascals), and electromagnetic waves (a form of transverse vacuum wave) express the amplitude in terms of its electric field (e.g., volts/meter).

The wavelength λ is the distance between two sequential crests or troughs (or other equivalent points), generally is measured in meters. A wavenumber k, the spatial frequency of the wave in radians per unit distance (typically per meter), can be associated with the wavelength by the relation

$$k = \frac{2\pi}{\lambda}.$$

The period T is the time for one complete cycle of an oscillation of a wave. The frequency f is the number of periods per unit time (per second) and is typically measured in hertz denoted as Hz. These are related by:

$$f = \frac{1}{T}.$$

In other words, the frequency and period of a wave are reciprocals.

The angular frequency ω represents the frequency in radians per second. It is related to the frequency or period by

$$\omega = 2\pi f = \frac{2\pi}{T}.$$

The wavelength λ of a sinusoidal waveform traveling at constant speed v is given by:

$$\lambda = \frac{v}{f},$$

where v is called the phase speed (magnitude of the phase velocity) of the wave and f is the wave's frequency.

Wavelength can be a useful concept even if the wave is not periodic in space. For example, in an ocean wave approaching shore, the incoming wave undulates with a varying *local* wavelength that depends in part on the depth of the sea floor compared to the wave height. The analysis of the wave can be based upon comparison of the local wavelength with the local water depth.

Although arbitrary wave shapes will propagate unchanged in lossless linear time-invariant systems, in the presence of dispersion the sine wave is the unique shape that will propagate unchanged but for phase and amplitude, making it easy to analyze. Due to the Kramers–Kronig relations, a linear medium with dispersion also exhibits loss, so the sine wave propagating in a dispersive medium is attenuated in certain frequency ranges that depend upon the medium. The sine function is periodic, so the sine wave or sinusoid has a wavelength in space and a period in time.

The sinusoid is defined for all times and distances, whereas in physical situations we usually deal with waves that exist for a limited span in space and duration in time. Fortunately, an arbitrary wave shape can be decomposed into an infinite set of sinusoidal waves by the use of Fourier analysis. As a result, the simple case of a single sinusoidal wave can be applied to more general cases. In particular, many media are linear, or nearly so, so the calculation of arbitrary wave behavior can be found by adding up responses to individual sinusoidal waves using the superposition principle to find the solution for a general waveform. When a medium is nonlinear, the response to complex waves cannot be determined from a sine-wave decomposition.

Standing Waves

Standing wave in stationary medium. The red dots represent the wave nodes

A standing wave, also known as a *stationary wave*, is a wave that remains in a constant position. This phenomenon can occur because the medium is moving in the opposite direction to the wave, or it can arise in a stationary medium as a result of interference between two waves traveling in opposite directions.

The *sum* of two counter-propagating waves (of equal amplitude and frequency) creates a *standing wave*. Standing waves commonly arise when a boundary blocks further propagation of the wave, thus causing wave reflection, and therefore introducing a counter-propagating wave. For example, when a violin string is displaced, transverse waves propagate out to where the string is held in place at the bridge and the nut, where the waves are reflected back. At the bridge and nut, the two opposed waves are in antiphase and cancel each other, producing a node. Halfway between two nodes there is an antinode, where the two counter-propagating waves *enhance* each other maximally. There is no net propagation of energy over time.

Physical Properties

Light beam exhibiting reflection, refraction, transmission and dispersion when encountering a prism

Waves exhibit common behaviors under a number of standard situations, e. g.

Transmission and Media

Waves normally move in a straight line (i.e. rectilinearly) through a *transmission medium*. Such media can be classified into one or more of the following categories:

- A *bounded medium* if it is finite in extent, otherwise an *unbounded medium*
- A *linear medium* if the amplitudes of different waves at any particular point in the medium can be added
- A *uniform medium* or *homogeneous medium* if its physical properties are unchanged at different locations in space
- An *anisotropic medium* if one or more of its physical properties differ in one or more directions
- An *isotropic medium* if its physical properties are the *same* in all directions

Absorption

Absorption of waves means, if a kind of wave strikes a matter, it will be absorbed by the matter. When a wave with that same natural frequency impinges upon an atom, then the electrons of that atom will be set into vibrational motion. If a wave of a given frequency strikes a material with electrons having the same vibrational frequencies, then those electrons will absorb the energy of the wave and transform it into vibrational motion.

Reflection

When a wave strikes a reflective surface, it changes direction, such that the angle made by the incident wave and line normal to the surface equals the angle made by the reflected wave and the same normal line.

Interference

Waves that encounter each other combine through superposition to create a new wave called an interference pattern. Important interference patterns occur for waves that are in phase.

Refraction

Sinusoidal traveling plane wave entering a region of lower wave velocity at an angle, illustrating the decrease in wavelength and change of direction (refraction) that results.

Refraction is the phenomenon of a wave changing its speed. Mathematically, this means that the size of the phase velocity changes. Typically, refraction occurs when a wave passes from one medium into another. The amount by which a wave is refracted by a material is given by the refractive index of the material. The directions of incidence and refraction are related to the refractive indices of the two materials by Snell's law.

Diffraction

A wave exhibits diffraction when it encounters an obstacle that bends the wave or when it spreads after emerging from an opening. Diffraction effects are more pronounced when the size of the obstacle or opening is comparable to the wavelength of the wave.

Polarization

The phenomenon of polarization arises when wave motion can occur simultaneously in two orthogonal directions. Transverse waves can be polarized, for instance. When polarization is used as a descriptor without qualification, it usually refers to the special, simple case of linear polarization. A transverse wave is linearly polarized if it oscillates in only one direction or plane. In the case of linear polarization. it is often useful to add the relative orientation of that plane, perpendicular to the direction of travel, in which the oscillation occurs, such as "horizontal" for instance, if the plane of polarization is parallel to the ground. Electromagnetic waves propagating in free space, for instance, are transverse; they can be polarized by the use of a polarizing filter.

Longitudinal waves, such as sound waves, do not exhibit polarization. For these waves there is only one direction of oscillation, that is, along the direction of travel.

Dispersion

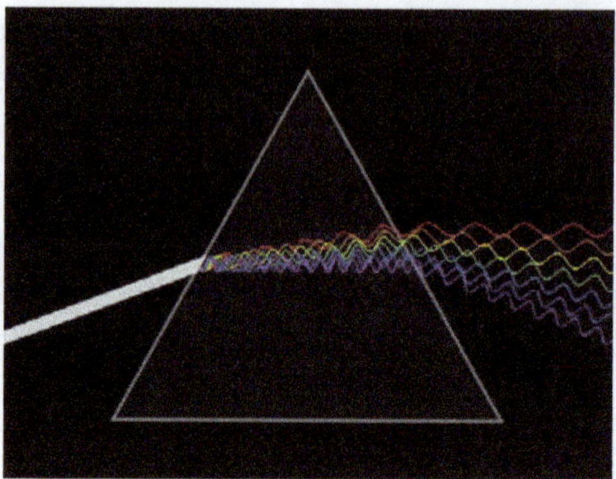
Schematic of light being dispersed by a prism.

A wave undergoes dispersion when either the phase velocity or the group velocity depends on the wave frequency. Dispersion is most easily seen by letting white light pass through a prism, the result of which is to produce the spectrum of colours of the rainbow. Isaac Newton performed experiments with light and prisms, presenting his findings in the *Opticks* (1704) that white light consists of several colours and that these colours cannot be decomposed any further.

Mechanical Waves

Waves on strings

The speed of a transverse wave traveling along a vibrating string (v) is directly proportional to the square root of the tension of the string (T) over the linear mass density (μ) :

$$v = \sqrt{\frac{T}{\mu}},$$

where the linear density μ is the mass per unit length of the string.

Acoustic Waves

Acoustic or sound waves travel at speed given by

$$v = \sqrt{\frac{B}{\rho_0}},$$

or the square root of the adiabatic bulk modulus divided by the ambient fluid density.

Water Waves

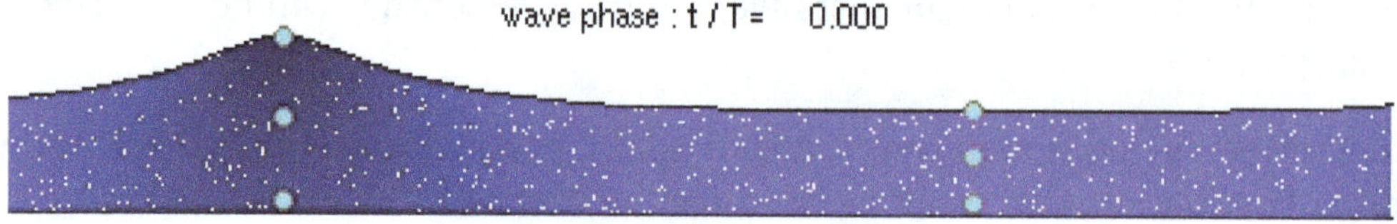

- Ripples on the surface of a pond are actually a combination of transverse and longitudinal waves; therefore, the points on the surface follow orbital paths.
- Sound—a mechanical wave that propagates through gases, liquids, solids and plasmas;
- Inertial waves, which occur in rotating fluids and are restored by the Coriolis effect;
- Ocean surface waves, which are perturbations that propagate through water.

Shock Waves

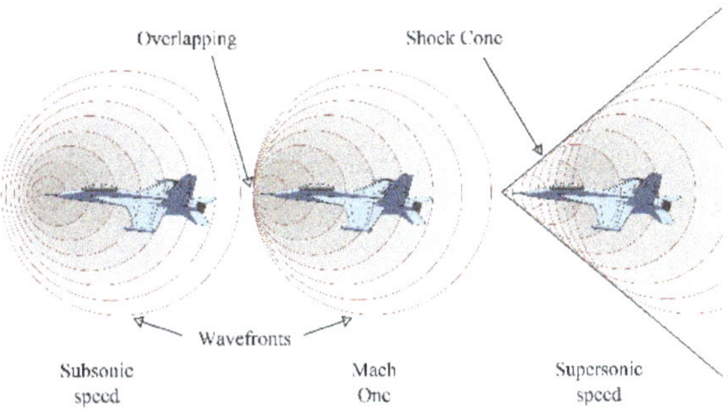

Other

- Waves of traffic, that is, propagation of different densities of motor vehicles, and so forth, which can be modeled as kinematic waves.
- Metachronal wave refers to the appearance of a traveling wave produced by coordinated sequential actions.
- It is worth noting that the mass-energy equivalence equation can be solved for this form:

$$c = \sqrt{\frac{e}{m}}.$$

Electromagnetic Waves

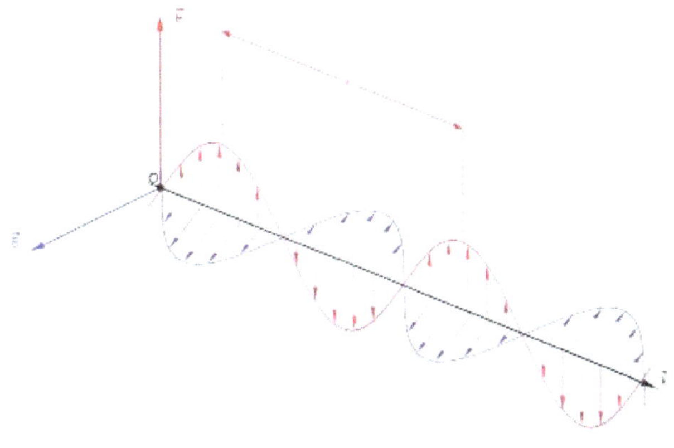

An electromagnetic wave consists of two waves that are oscillations of the electric and magnetic fields. An electromagnetic wave travels in a direction that is at right angles to the oscillation direction of both fields. In the 19th century, James Clerk Maxwell showed that, in vacuum, the electric and magnetic fields satisfy the wave equation both with speed equal to that of the speed of light. From this emerged the idea that light is an electromagnetic wave. Electromagnetic waves can have different frequencies (and thus wavelengths), giving rise to various types of radiation such as radio waves, microwaves, infrared, visible light, ultraviolet, X-rays, and Gamma rays.

Quantum Mechanical Waves

Schrödinger Equation

The Schrödinger equation describes the wave-like behavior of particles in quantum mechanics. Solutions of this equation are wave functions which can be used to describe the probability density of a particle.

Dirac Equation

The Dirac equation is a relativistic wave equation detailing electromagnetic interactions. Dirac waves accounted for the fine details of the hydrogen spectrum in a completely rigorous way. The wave equation also implied the existence of a new form of matter, antimatter, previously unsuspected and unobserved and which was experimentally confirmed. In the context of quantum field theory, the Dirac equation is reinterpreted to describe quantum fields corresponding to spin-½ particles.

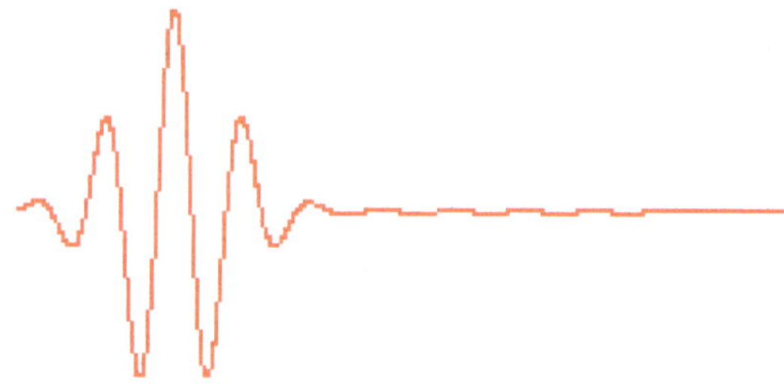

A propagating wave packet; in general, the *envelope* of the wave packet moves at a different speed than the constituent waves.

De Broglie Waves

Louis de Broglie postulated that all particles with momentum have a wavelength

$$\lambda = \frac{h}{p},$$

where h is Planck's constant, and p is the magnitude of the momentum of the particle. This hypothesis was at the basis of quantum mechanics. Nowadays, this wavelength is called the de Broglie wavelength. For example, the electrons in a CRT display have a de Broglie wavelength of about 10^{-13} m.

A wave representing such a particle traveling in the k-direction is expressed by the wave function as follows:

$$\psi(\mathbf{r}, t = 0) = A\, e^{i\mathbf{k}\cdot\mathbf{r}} \,,$$

where the wavelength is determined by the wave vector k as:

$$\lambda = \frac{2\pi}{k} \,,$$

and the momentum by:

$$\mathbf{p} = \hbar\mathbf{k} \,.$$

However, a wave like this with definite wavelength is not localized in space, and so cannot represent a particle localized in space. To localize a particle, de Broglie proposed a superposition of different wavelengths ranging around a central value in a wave packet, a waveform often used in quantum mechanics to describe the wave function of a particle. In a wave packet, the wavelength of the particle is not precise, and the local wavelength deviates on either side of the main wavelength value.

In representing the wave function of a localized particle, the wave packet is often taken to have a Gaussian shape and is called a *Gaussian wave packet*. Gaussian wave packets also are used to analyze water waves.

For example, a Gaussian wavefunction ψ might take the form:

$$\psi(x, t = 0) = A \exp\left(-\frac{x^2}{2\sigma^2} + ik_0 x \right),$$

at some initial time $t = 0$, where the central wavelength is related to the central wave vector k_0 as $\lambda_0 = 2\pi / k_0$. It is well known from the theory of Fourier analysis, or from the Heisenberg uncertainty principle (in the case of quantum mechanics) that a narrow range of wavelengths is necessary to produce a localized wave packet, and the more localized the envelope, the larger the spread in required wavelengths. The Fourier transform of a Gaussian is itself a Gaussian. Given the Gaussian:

$$f(x) = e^{-x^2/(2\sigma^2)} \,,$$

the Fourier transform is:

$$\tilde{f}(k) = \sigma e^{-\sigma^2 k^2/2} \,.$$

The Gaussian in space therefore is made up of waves:

$$f(x) = \frac{1}{\sqrt{2\pi}} \int_{-\infty}^{\infty} \tilde{f}(k) e^{ikx}\, dk \,;$$

that is, a number of waves of wavelengths λ such that $k\lambda = 2\Pi$.

The parameter σ decides the spatial spread of the Gaussian along the *x*-axis, while the Fourier transform shows a spread in wave vector *k* determined by 1/σ. That is, the smaller the extent in space, the larger the extent in *k*, and hence in λ = 2Π/*k*.

Gravity Waves

Gravity waves are waves generated in a fluid medium or at the interface between two media when the force of gravity or buoyancy tries to restore equilibrium. A ripple on a pond is one example.

Gravitational Waves

Gravitational waves also travel through space. The first observation of gravitational waves was announced on 11 February 2016. Gravitational waves are disturbances in the curvature of spacetime, predicted by Einstein's theory of general relativity.

WKB Method

In a nonuniform medium, in which the wavenumber *k* can depend on the location as well as the frequency, the phase term *kx* is typically replaced by the integral of *k*(*x*)*dx*, according to the WKB method. Such nonuniform traveling waves are common in many physical problems, including the mechanics of the cochlea and waves on hanging ropes.

We shift our attention to oscillations that propagate in space as time evolves. This is referred to as a wave. The sinusoidal wave

$$a(x,t) = A\cos(\omega t - kx + \psi),$$

is the simplest example of a wave, we shall consider other possibilities later in the course. It is often convenient to represent the wave in the complex notation introduced earlier. We have

$$\tilde{a}(x,t) = \tilde{A}e^{i(\omega t - kx)}.$$

What is *a(x,t)* ?

The wave phenomena is found in many different situations, and $a(x, t)$ represents a different physical quantity in each situation. For example, it is well known that disturbances in air propagate from one point to another as waves and are perceived by us as sound. Any source of sound (eg. a loud speaker) produces compressions and rarefactions in the air, and the patterns of compressions and rarefactions propagate from one point to another. Using $\rho(x, t)$ to denote the air density, we can express this as $\rho(x,t) = \bar{\rho} + \Delta\rho(x,t)$ where $\bar{\rho}$ is the density in the absence of the disturbance and $\Delta\rho(x,t)$ is the change due to the disturbance. We can use equation to represent a sinusoidal sound wave if we identify $a(x, t)$ with $\Delta\rho(x,t)$.

The transverse vibrations of a stretched string is another examples. In this situation $a(x, t)$ corresponds to $y(x, t)$ which is the displacement of the string.

Angular Frequency and Wave Number

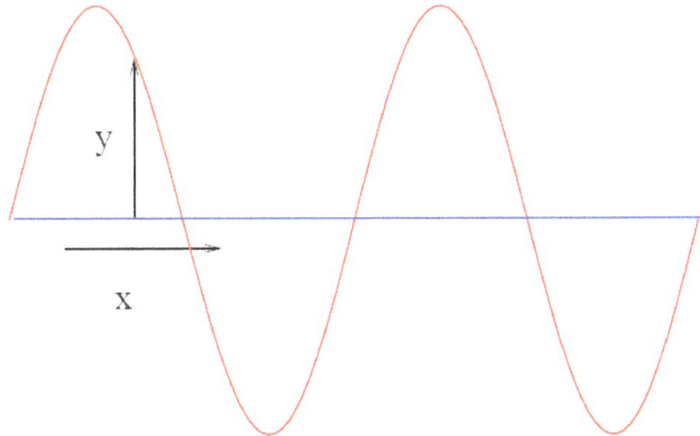

Transverse vibrations of a stretched string.

The sinusoidal wave in equation has a complex amplitude $\tilde{A} = Ae^{i\psi}$. Here A, the magnitude of $\tilde{A}$ determines the magnitude of the wave. We refer

to $\varphi(x, t) = \omega t - kx + \psi$ as the phase of the wave, and the wave can be also expressed as

$$\tilde{a}(x,t) = Ae^{i\phi(x,t)},$$

If we study the behaviour of the wave at a fixed position x_1, we have

$$\tilde{a}(t) = \left[\tilde{A}e^{-ikx_1}\right]e^{i\omega t} = \tilde{A}'e^{i\omega t}.$$

We see that this is the familiar oscillation (SHO). The oscillation has amplitude $\tilde{A}' = [\tilde{A}e^{-ikx_1}]$ which includes an extra constant phase factor. The value of a(t) has sinusoidal variations. Starting at t = 0, the behaviour repeats after a time period T when $\omega T = 2\pi$. We identify ω as the angular frequency of the wave related to the frequency v as

$$\omega = \frac{2\pi}{T}2\pi v.$$

We next study the wave as a function of position x at a fixed instant of time t_1. We have

$$\tilde{a}(x) = \tilde{\mathrm{A}}\mathrm{e}^{i\omega t_1}e^{-ikx} = \tilde{\mathrm{A}}''\mathrm{e}^{-ikx},$$

where we have absorbed the extra phase $e^{i\omega t}$ in the complex amplitude $\tilde{\mathrm{A}}''$. This tells us that the spatial variation is also sinusoidal as shown in Figure. The wavelength λ is the distance after which $a(x)$ repeats itself. Starting from $x = 0$, we see that $a(x)$ repeats when $kx = 2\pi$ which tells us that $k\lambda = 2\pi$ or

$$\mathrm{k} = \frac{2\pi}{\lambda},$$

where we refer to k as the wave number. We note that the wave number and the angular frequency tell us the rate of change of the phase $\phi(x, t)$ with position and time respectively

$$k=-\frac{\partial\phi}{\partial x} \textit{ and } \omega=\frac{\partial\phi}{\partial t}.$$

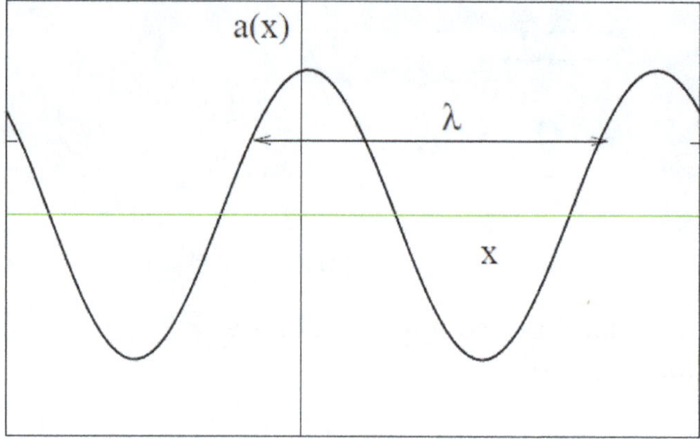

Variation of wave with space for a fixed time.

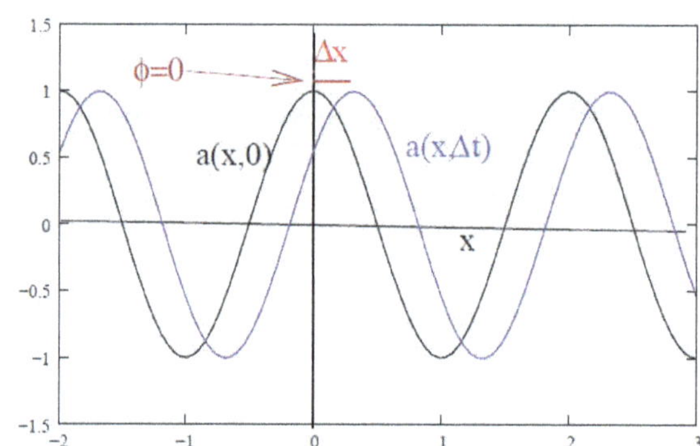

The movement of constant phase $\phi = 0$,as function of time.

Phase Velocity

We now consider the evolution of the wave in both position and time together. We consider the wave

$$\tilde{a}(x,t) = Ae^{i(\omega t-kx)},$$

which has phase $\phi(x, t) = \omega t - kx$. Let us follow the motion of the position where the phase has value $\phi(x, t) = 0$ as time increases. We see that initially $\phi = 0$ at $x = 0, t = 0$ and after a time Δt this moves to a position

$$\Delta x = \left(\frac{\omega}{k}\right)\Delta t,$$

shown in Figure. The point with phase $\phi = 0$ moves at speed

$$\upsilon_p = \left(\frac{\omega}{k}\right).$$

It is not difficult to convince oneself that this is true for any constant value of the phase, and the whole sinusoidal pattern propagates along the +x direction at the speed υ_p which is called the phase velocity of the wave.

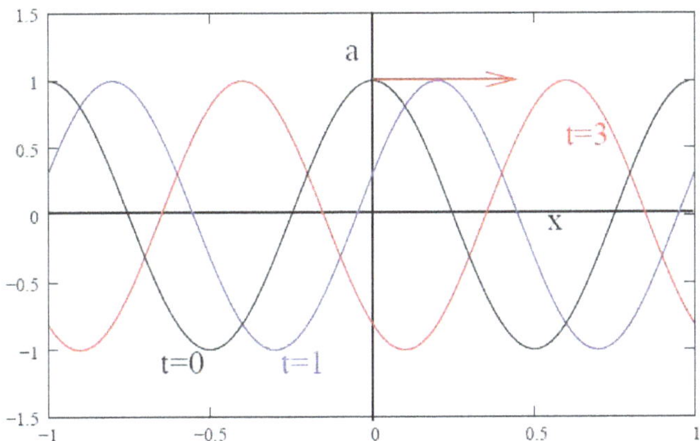

Propagation of wave along the x-direction.

Waves in Three Dimensions

We have till now considered waves which depend on only one position coordinate x and time t. This is quite adequate when considering waves on a string as the position along a string can be described by a single coordinate. It is necessary to bring three spatial coordinates (x, y, z) into the picture when considering a wave propagating in three dimensional space. A sound wave propagating in air is an example.

We use the vector $\vec{r} = x\hat{i} + y\hat{j} + z\hat{k}$ to denote a point in three dimensional space. The solution which we have been discussing

$$\tilde{a}\left(\vec{r},t\right) = Ae^{i(\omega t - kx)},$$

can be interpreted in the context of a three dimensional space. Note that $\tilde{a}\left(\vec{r},t\right)$ varies only along the x direction and not along y and z. Considering the phase $\phi\left(\vec{r},t\right) = \omega t - kx$ we see that at any particular instant of time t, there are surfaces on which the phase is constant. The constant phase surfaces of a wave are called wave fronts. In this case the wave fronts are parallel to the $y-z$ plane as shown in Figure. The wave fronts move along the $+x$ direction with speed υ_p as time evolves. You can check this by following the motion of the $\phi = 0$ surface shown in the following Figure.

Waves in an Arbitrary Direction

Let us now discuss how to describe a sinusoidal plane wave in an arbitrary direction denoted by the unit vector $\vec{n}$. A wave propagating along the $\hat{i}$ direction can be written as

$$\tilde{a}(\vec{r},t) = \tilde{A}e^{i(\omega t - \vec{k}\cdot\vec{r})},$$

where $\vec{k} = k\hat{i}$ is called the wave vector. Note that $\vec{k}$ is different from $\hat{k}$ which is the unit vector along the z direction. It is now obvious that a wave along an arbitrary direction $\hat{n}$ can also be represented by eq. if we change the

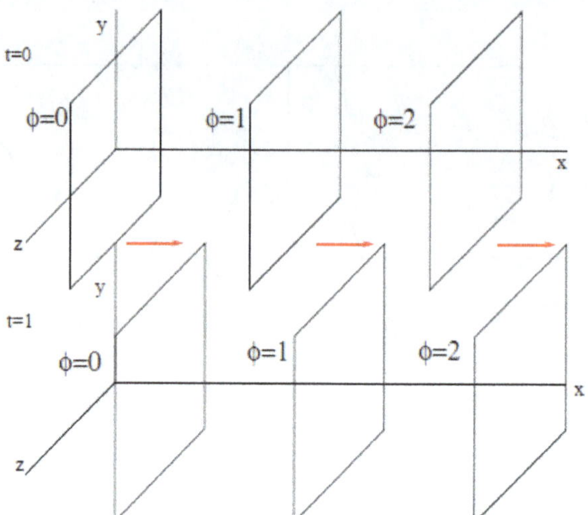

Wavefronts for a wave propagating along the x direction.

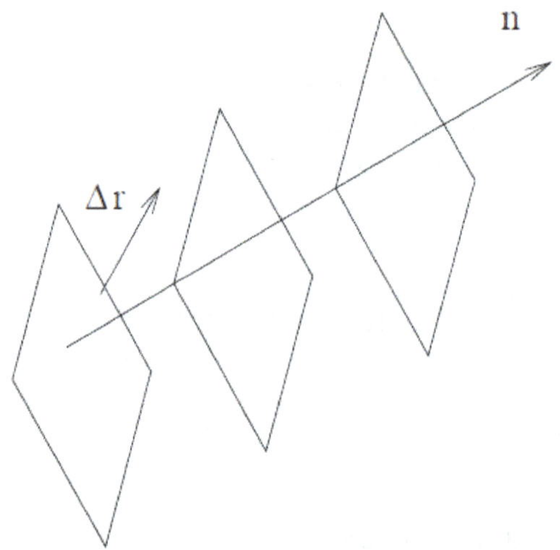

Wavefronts for a wave propagating along the direction $\hat{n}$.

wave vector to $\vec{k} = k\hat{n}$. The wave vector $\vec{k}$ carries information about both the wavelength λ and the direction of propagation $\hat{n}$.

For such a wave, at a fixed instant of time, the phase $\phi(\vec{r},t) = \omega t - \vec{k}\cdot\vec{r}$ changes only along $\hat{n}$. The wave fronts are surfaces perpendicular to $\hat{n}$ as shown in Figure.

References

- Michael H. Tooley (2006). Electronic Circuits: Fundamentals and Applications. Newnes. pp. 77–78. ISBN 978-0-7506-6923-8

- Snyder; Farley (2011). "Energetically optimal stride frequency in running: the effects of incline and decline". The Journal of Experimental Biology. 214: 2089–95. PMID 21613526. doi:10.1242/jeb.053157. Retrieved 1 Sep 2014
- Lev A. Ostrovsky & Alexander I. Potapov (2002). Modulated waves: theory and application. Johns Hopkins University Press. ISBN 0-8018-7325-8
- K. Yusuf Billah and Robert H. Scanlan (1991). "Resonance, Tacoma Narrows Bridge Failure, and Undergraduate Physics Textbooks" (PDF). American Journal of Physics. 59 (2): 118–124. Bibcode:1991AmJPh..59..118B. doi:10.1119/1.16590. Retrieved 2011-05-29
- Michael A. Slawinski (2003). "Wave equations". Seismic waves and rays in elastic media. Elsevier. pp. 131 ff. ISBN 0-08-043930-6
- For an example derivation, see the steps leading up to eq. (17) in Francis Redfern. "Kinematic Derivation of the Wave Equation". Physics Journal

3

Electromagnetic Radiation and Waves

Electromagnetic radiations are self-propagating waves of electromagnetic field containing electromagnetic energy. It is possible for electromagnetic waves to oscillate with the help of an electric dipole. This concept is very important to manufacture devices like the laser. The chapter strategically encompasses and incorporates the major components and key concepts of electromagnetic radiation and electromagnetic waves, providing a complete understanding.

Electromagnetic Radiation

In physics, electromagnetic radiation (EM radiation or EMR) refers to the waves (or their quanta, photons) of the electromagnetic field, propagating (radiating) through space carrying electromagnetic radiant energy. It includes radio waves, microwaves, infrared, (visible) light, ultraviolet, X-, and gamma radiation.

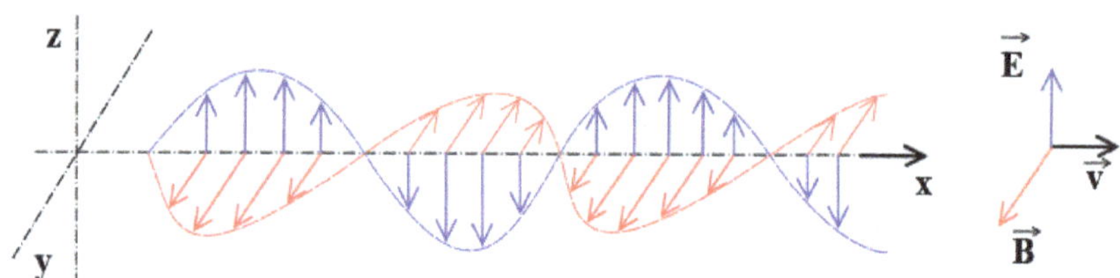

The electromagnetic waves that compose electromagnetic radiation can be imagined as a self-propagating transverse oscillating wave of electric and magnetic fields. This diagram shows a plane linearly polarized EMR wave propagating from left to right (X axis). The electric field is in a vertical plane (Z axis) and the magnetic field in a horizontal plane (Y axis). The electric and magnetic fields in EMR waves are always in phase and at 90 degrees to each other.

Classically, electromagnetic radiation consists of electromagnetic waves, which are synchronized oscillations of electric and magnetic fields that propagate at the speed of light through a vacuum. The oscillations of the two fields are perpendicular to each other and perpendicular to the direction of energy and wave propagation, forming a transverse wave. The wavefront of electromagnetic waves emitted from a point source (such as a lightbulb) is a sphere. The position of an electromagnetic wave within the electromagnetic spectrum can be characterized by either its frequency of oscillation or its wavelength. The electromagnetic spectrum includes, in order of increasing frequency and decreasing wavelength: radio waves, microwaves, infrared radiation, visible light, ultraviolet radiation, X-rays and gamma rays.

Electromagnetic waves are produced whenever charged particles are accelerated, and these waves can subsequently interact with other charged particles. EM waves carry energy, momentum and angular momentum away from their source particle and can impart those quantities to matter with which they interact. Quanta of EM waves are called photons, whose rest mass is zero, but whose energy, or equivalent total (relativistic) mass, is not zero so they are still affected by gravity. Electromagnetic radiation is associated with those EM waves that are free to propagate themselves ("radiate") without the continuing influence of the moving charges that produced them, because they have achieved sufficient distance from those charges. Thus, EMR is sometimes referred to as the far field. In this language, the *near field* refers to EM fields near the charges and current that directly produced them, specifically, electromagnetic induction and electrostatic induction phenomena.

In the quantum theory of electromagnetism, EMR consists of photons, the elementary particles responsible for all electromagnetic interactions. Quantum effects provide additional sources of EMR, such as the transition of electrons to lower energy levels in an atom and black-body radiation. The energy of an individual photon is quantized and is greater for photons of higher frequency. This relationship is given by Planck's equation $E = hv$, where E is the energy per photon, v is the frequency of the photon, and h is Planck's constant. A single gamma ray photon, for example, might carry ~100,000 times the energy of a single photon of visible light.

The effects of EMR upon chemical compounds and biological organisms depend both upon the radiation's power and its frequency. EMR of visible or lower frequencies (i.e., visible light, infrared, microwaves, and radio waves) is called *non-ionizing radiation*, because its photons do not individually have enough energy to ionize atoms or molecules. The effects of these radiations on chemical systems and living tissue are caused primarily by heating effects from the combined energy transfer of many photons. In contrast, high ultraviolet, X-rays and gamma rays are called *ionizing radiation* since individual photons of such high frequency have enough energy to ionize molecules or break chemical bonds. These radiations have the ability to cause chemical reactions and damage living cells beyond that resulting from simple heating, and can be a health hazard.

Theory

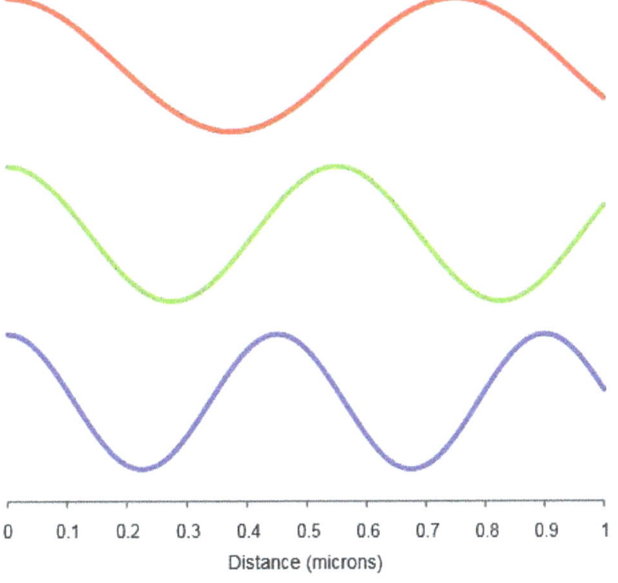

Shows the relative wavelengths of the electromagnetic waves of three different colours of light (blue, green, and red) with a distance scale in micrometers along the x-axis.

Maxwell's Equations

Maxwell derived a wave form of the electric and magnetic equations, thus uncovering the wave-like nature of electric and magnetic fields and their symmetry. Because the speed of EM waves predicted by the wave equation coincided with the measured speed of light, Maxwell concluded that light itself is an EM wave. Maxwell's equations were confirmed by Heinrich Hertz through experiments with radio waves.

According to Maxwell's equations, a spatially varying electric field is always associated with a magnetic field that changes over time. Likewise, a spatially varying magnetic field is associated with specific changes over time in the electric field. In an electromagnetic wave, the changes in the electric field are always accompanied by a wave in the magnetic field in one direction, and vice versa. This relationship between the two occurs without either type field causing the other; rather, they occur together in the same way that time and space changes occur together and are interlinked in special relativity. In fact, magnetic fields may be viewed as relativistic distortions of electric fields, so the close relationship between space and time changes here is more than an analogy. Together, these fields form a propagating electromagnetic wave, which moves out into space and need never again affect the source. The distant EM field formed in this way by the acceleration of a charge carries energy with it that "radiates" away through space, hence the term.

Near and Far Fields

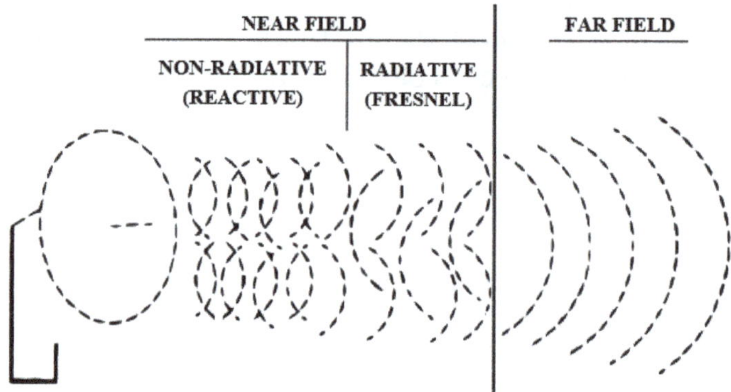

In electromagnetic radiation (such as microwaves from an antenna, shown here) the term applies only to the parts of the electromagnetic field that radiate into infinite space and decrease in intensity by an inverse-square law of power, so that the total radiation energy that crosses through an imaginary spherical surface is the same, no matter how far away from the antenna the spherical surface is drawn. Electromagnetic radiation thus includes the far field part of the electromagnetic field around a transmitter. A part of the "near-field" close to the transmitter, forms part of the changing electromagnetic field, but does not count as electromagnetic radiation.

Maxwell's equations established that some charges and currents ("sources") produce a local type of electromagnetic field near them that does *not* have the behaviour of EMR. Currents directly produce a magnetic field, but it is of a magnetic dipole type that dies out with distance from the current. In a similar manner, moving charges pushed apart in a conductor by a changing electrical potential (such as in an antenna) produce an electric dipole type electrical field, but this also declines with distance. These fields make up the near-field near the EMR source. Neither of these behaviours are responsible for EM radiation. Instead, they cause electromagnetic field behaviour that only efficiently transfers power to a receiver very close to the source, such as the magnetic induction inside a transformer, or the feedback behaviour that happens close to the coil of a metal detector. Typically, near-fields have a powerful effect on their own sources, causing an increased "load" (decreased electrical reactance) in the source or transmitter, whenever energy is withdrawn from the EM field by a receiver. Otherwise, these fields do not "propagate" freely out into space, carrying their energy away without distance-limit, but rather oscillate, returning their energy to the transmitter if it is not received by a receiver.

By contrast, the EM far-field is composed of *radiation* that is free of the transmitter in the sense that (unlike the case in an electrical transformer) the transmitter requires the same power to send these changes in the fields out, whether the signal is immediately picked up or not. This distant part of the electromagnetic field *is* "electromagnetic radiation" (also called the far-field). The far-fields propagate (radiate) without allowing the transmitter to affect them. This causes them to be independent in the sense that their existence and their energy, after they have left the transmitter, is completely independent of both transmitter and receiver. Because such waves conserve the amount of energy they transmit through any spherical boundary surface drawn around their source, and because such surfaces have an area that is defined by the square of the distance from the source, the power of EM radiation always varies according to an inverse-square law. This is in contrast to dipole parts of the EM field close to the source (the near-field), which varies in power according to an inverse cube power law, and thus does *not* transport a conserved amount of energy over distances, but instead fades with distance, with its energy (as noted) rapidly returning to the transmitter or absorbed by a nearby receiver (such as a transformer secondary coil).

The far-field (EMR) depends on a different mechanism for its production than the near-field, and upon different terms in Maxwell's equations. Whereas the magnetic part of the near-field is due to currents in the source, the magnetic field in EMR is due only to the local change in the electric field. In a similar way, while the electric field in the near-field is due directly to the charges and charge-separation in the source, the electric field in EMR is due to a change in the local magnetic field. Both processes for producing electric and magnetic EMR fields have a different dependence on distance than do near-field dipole electric and magnetic fields. That is why the EMR type of EM field becomes dominant in power "far" from sources. The term "far from sources" refers to how far from the source (moving at the speed of light) any portion of the outward-moving EM field is located, by the time that source currents are changed by the varying source potential, and the source has therefore begun to generate an outwardly moving EM field of a different phase.

A more compact view of EMR is that the far-field that composes EMR is generally that part of the EM field that has traveled sufficient distance from the source, that it has become completely disconnected from any feedback to the charges and currents that were originally responsible for it. Now independent of the source charges, the EM field, as it moves farther away, is dependent only upon the accelerations of the charges that produced it. It no longer has a strong connection to the direct fields of the charges, or to the velocity of the charges (currents).

In the Liénard–Wiechert potential formulation of the electric and magnetic fields due to motion of a single particle (according to Maxwell's equations), the terms associated with acceleration of the particle are those that are responsible for the part of the field that is regarded as electromagnetic radiation. By contrast, the term associated with the changing static electric field of the particle and the magnetic term that results from the particle's uniform velocity, are both associated with the electromagnetic near-field, and do not comprise EM radiation.

Properties

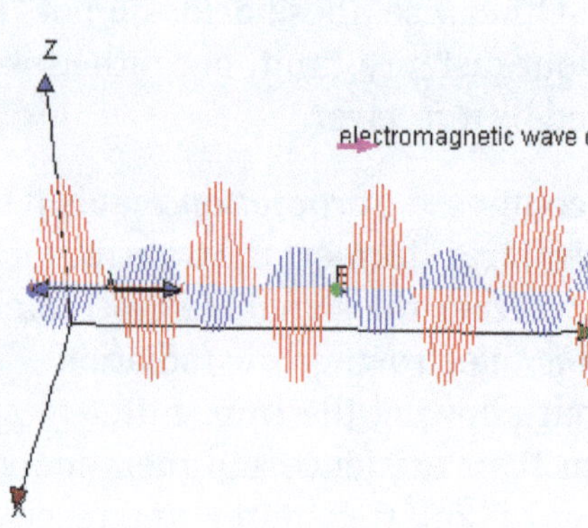

Electromagnetic waves can be imagined as a self-propagating transverse oscillating wave of electric and magnetic fields. This 3D animation shows a plane linearly polarized wave propagating from left to right. Note that the electric and magnetic fields in such a wave are in-phase with each other, reaching minima and maxima together

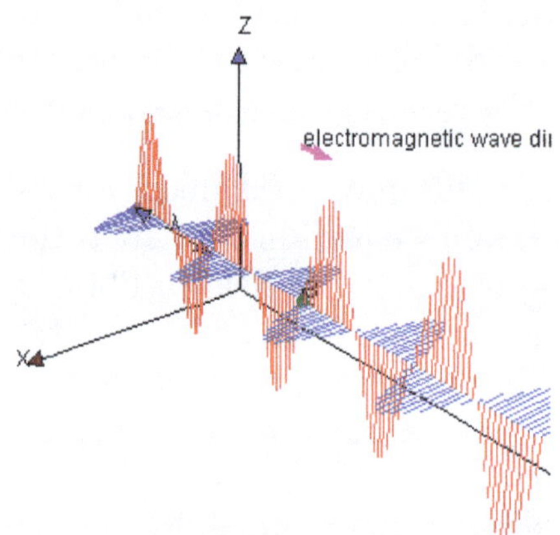

An alternate view of the wave shown above.

Electrodynamics is the physics of electromagnetic radiation, and electromagnetism is the physical phenomenon associated with the theory of electrodynamics. Electric and magnetic fields obey the properties of superposition. Thus, a field due to any particular particle or time-varying electric or magnetic field contributes to the fields present in the same space due to other causes. Further, as they are vector fields, all magnetic and electric field vectors add together according to vector addition. For example, in optics two or more coherent lightwaves may interact and by constructive or destructive interference yield a resultant irradiance deviating from the sum of the component irradiances of the individual lightwaves.

Since light is an oscillation it is not affected by traveling through static electric or magnetic fields in a linear medium such as a vacuum. However, in nonlinear media, such as some crystals, interactions can occur between light and static electric and magnetic fields — these interactions include the Faraday effect and the Kerr effect.

In refraction, a wave crossing from one medium to another of different density alters its speed and direction upon entering the new medium. The ratio of the refractive indices of the media determines the degree of refraction, and is summarized by Snell's law. Light of composite wavelengths (natural sunlight) disperses into a visible spectrum passing through a prism, because of the wavelength-dependent refractive index of the prism material (dispersion); that is, each component wave within the composite light is bent a different amount.

EM radiation exhibits both wave properties and particle properties at the same time. Both wave and particle characteristics have been confirmed in many experiments. Wave characteristics are more apparent when EM radiation is measured over relatively large timescales and over large distances while particle characteristics are more evident when measuring small timescales and distances. For example, when electromagnetic radiation is absorbed by matter, particle-like properties will be more obvious when the average number of photons in the cube of the relevant wavelength is much smaller than 1. It is not too difficult to experimentally observe non-uniform deposition of energy when light is absorbed, however this alone is not evidence of "particulate" behavior. Rather, it reflects the quantum nature of *matter*. Demonstrating that the light itself is quantized, not merely its interaction with matter, is a more subtle affair.

Some experiments display both the wave and particle natures of electromagnetic waves, such as the self-interference of a single photon. When a single photon is sent through an interferometer, it passes through both paths, interfering with itself, as waves do, yet is detected by a photomultiplier or other sensitive detector only once.

A quantum theory of the interaction between electromagnetic radiation and matter such as electrons is described by the theory of quantum electrodynamics.

Electromagnetic waves can be polarized, reflected, refracted, diffracted or interfere with each other.

Wave Model

Electromagnetic radiation is a transverse wave, meaning that its oscillations are perpendicular to the direction of energy transfer and travel. The electric and magnetic parts of the field stand in a fixed ratio of strengths in order to satisfy the two Maxwell equations that specify how one is produced from the other. These E and B fields are also in phase, with both reaching maxima and minima at the same points in space. A common misconception is that the E and B fields in electromagnetic radiation are out of phase because a change in one produces the other, and this would produce a phase difference between them as sinusoidal functions (as indeed happens in electromagnetic induction, and in the near-field close to antennas). However, in the far-field EM radiation which is described by the two source-free Maxwell curl operator equations, a more correct description is that a time-change in one type of field is proportional to a space-change in the other. These derivatives require that the E and B fields in EMR are in-phase.

An important aspect of light's nature is its frequency. The frequency of a wave is its rate of oscillation and is measured in hertz, the SI unit of frequency, where one hertz is equal to one oscillation per second. Light usually has multiple frequencies that sum to form the resultant wave. Different frequencies undergo different angles of refraction, a phenomenon known as dispersion.

A wave consists of successive troughs and crests, and the distance between two adjacent crests or troughs is called the wavelength. Waves of the electromagnetic spectrum vary in size, from very long radio waves the size of buildings to very short gamma rays smaller than atom nuclei. Frequency is inversely proportional to wavelength, according to the equation:

$$v = f\lambda$$

where v is the speed of the wave (c in a vacuum, or less in other media), f is the frequency and λ is the wavelength. As waves cross boundaries between different media, their speeds change but their frequencies remain constant.

Electromagnetic waves in free space must be solutions of Maxwell's electromagnetic wave equation. Two main classes of solutions are known, namely plane waves and spherical waves. The plane waves may be viewed as the limiting case of spherical waves at a very large (ideally infinite) distance from the source. Both types of waves can have a waveform which is an arbitrary time function (so long as it is sufficiently differentiable to conform to the wave equation). As with any time function, this can be decomposed by means of Fourier analysis into its frequency spectrum, or individual sinusoidal components, each of which contains a single frequency, amplitude and phase. Such a component wave is said to be *monochromatic*. A monochromatic electromagnetic wave can be characterized by its frequency or wavelength, its peak amplitude, its phase relative to some reference phase, its direction of propagation and its polarization.

Interference is the superposition of two or more waves resulting in a new wave pattern. If the fields have components in the same direction, they constructively interfere, while opposite directions cause destructive interference. An example of interference caused by EMR is electromagnetic interference (EMI) or as it is more commonly known as, radio-frequency interference (RFI). Additionally, multiple polarization signals can be combined (i.e. interfered) to form new states of polarization, which is known as parallel polarization state generation.

The energy in electromagnetic waves is sometimes called radiant energy.

Particle Model and Quantum Theory

An anomaly arose in the late 19th century involving a contradiction between the wave theory of light and measurements of the electromagnetic spectra that were being emitted by thermal radiators known as black bodies. Physicists struggled with this problem, which later became known as the ultraviolet catastrophe, unsuccessfully for many years. In 1900, Max Planck developed a new theory of black-body radiation that explained the observed spectrum. Planck's theory was based on the idea that black bodies emit light (and other electromagnetic radiation) only as discrete bundles or packets of energy. These packets were called quanta. Later, Albert Einstein proposed that light quanta be regarded as real particles. Later the particle of light was given the name photon, to correspond with other particles being described around this time, such as the electron and proton. A photon has an energy, E, proportional to its frequency, f, by

$$E = hf = \frac{hc}{\lambda}$$

where h is Planck's constant, λ is the wavelength and c is the speed of light. This is sometimes known as the Planck–Einstein equation. In quantum theory the energy of the photons is thus directly proportional to the frequency of the EMR wave.

Likewise, the momentum p of a photon is also proportional to its frequency and inversely proportional to its wavelength:

$$p=\frac{E}{c}=\frac{hf}{c}=\frac{h}{\lambda}.$$

The source of Einstein's proposal that light was composed of particles (or could act as particles in some circumstances) was an experimental anomaly not explained by the wave theory: the photoelectric effect, in which light striking a metal surface ejected electrons from the surface, causing an electric current to flow across an applied voltage. Experimental measurements demonstrated that the energy of individual ejected electrons was proportional to the *frequency*, rather than the *intensity*, of the light. Furthermore, below a certain minimum frequency, which depended on the particular metal, no current would flow regardless of the intensity. These observations appeared to contradict the wave theory, and for years physicists tried in vain to find an explanation. In 1905, Einstein explained this puzzle by resurrecting the particle theory of light to explain the observed effect. Because of the preponderance of evidence in favor of the wave theory, however, Einstein's ideas were met initially with great skepticism among established physicists. Eventually Einstein's explanation was accepted as new particle-like behavior of light was observed, such as the Compton effect.

As a photon is absorbed by an atom, it excites the atom, elevating an electron to a higher energy level (one that is on average farther from the nucleus). When an electron in an excited molecule or atom descends to a lower energy level, it emits a photon of light at a frequency corresponding to the energy difference. Since the energy levels of electrons in atoms are discrete, each element and each molecule emits and absorbs its own characteristic frequencies. Immediate photon emission is called fluorescence, a type of photoluminescence. An example is visible light emitted from fluorescent paints, in response to ultraviolet (blacklight). Many other fluorescent emissions are known in spectral bands other than visible light. Delayed emission is called phosphorescence.

Wave–particle Duality

The modern theory that explains the nature of light includes the notion of wave–particle duality. More generally, the theory states that everything has both a particle nature and a wave nature, and various experiments can be done to bring out one or the other. The particle nature is more easily discerned using an object with a large mass. A bold proposition by Louis de Broglie in 1924 led the scientific community to realize that electrons also exhibited wave–particle duality.

Wave and Particle Effects of Electromagnetic Radiation

Together, wave and particle effects fully explain the emission and absorption spectra of EM radiation. The matter-composition of the medium through which the light travels determines the nature of the absorption and emission spectrum. These bands correspond to the allowed energy levels in the atoms. Dark bands in the absorption spectrum are due to the atoms in an intervening medium between source and observer. The atoms absorb certain frequencies of the light between emitter and detector/eye, then emit them in all directions. A dark band appears to the detector, due to the radiation scattered out of the beam. For instance, dark bands in the light emitted by a distant star are due to the atoms in the star's atmosphere. A similar phenomenon occurs for emission, which is seen when an emitting gas glows due to excitation of the atoms from any mechanism, including heat. As electrons descend to lower energy levels, a spectrum is emitted that represents the jumps between the energy levels of the electrons, but lines are seen because again emission happens only at particular energies after excitation. An example is the emission spectrum of nebulae. Rapidly moving electrons are most sharply accelerated when they encounter a region of force, so they are responsible for producing much of the highest frequency electromagnetic radiation observed in nature.

These phenomena can aid various chemical determinations for the composition of gases lit from behind (absorption spectra) and for glowing gases (emission spectra). Spectroscopy (for example) determines what chemical elements comprise a particular star. Spectroscopy is also used in the determination of the distance of a star, using the red shift.

Propagation Speed

Any electric charge that accelerates, or any changing magnetic field, produces electromagnetic radiation. Electromagnetic information about the charge travels at the speed of light. Accurate treatment thus incorporates a concept known as retarded time, which adds to the expressions for the electrodynamic electric field and magnetic field. These extra terms are responsible for electromagnetic radiation.

When any wire (or other conducting object such as an antenna) conducts alternating current, electromagnetic radiation is propagated at the same frequency as the current. In many such situations it is possible to identify an electrical dipole moment that arises from separation of charges due to the exciting electrical potential, and this dipole moment oscillates in time, as the charges move back and forth. This oscillation at a given frequency gives rise to changing electric and magnetic fields, which then set the electromagnetic radiation in motion.

At the quantum level, electromagnetic radiation is produced when the wavepacket of a charged particle oscillates or otherwise accelerates. Charged particles in a stationary state do not move, but a superposition of such states may result in a transition state that has an electric dipole moment that oscillates in time. This oscillating dipole moment is responsible for the phenomenon of radiative transition between quantum states of a charged particle. Such states occur (for example) in atoms when photons are radiated as the atom shifts from one stationary state to another.

As a wave, light is characterized by a velocity (the speed of light), wavelength, and frequency. As particles, light is a stream of photons. Each has an energy related to the frequency of the wave given by Planck's relation $E = hf$, where E is the energy of the photon, $h = 6.626 \times 10^{-34}$ J·s is Planck's constant, and f is the frequency of the wave.

One rule is obeyed regardless of circumstances: EM radiation in a vacuum travels at the speed of light, *relative to the observer*, regardless of the observer's velocity. (This observation led to Einstein's development of the theory of special relativity.)

In a medium (other than vacuum), velocity factor or refractive index are considered, depending on frequency and application. Both of these are ratios of the speed in a medium to speed in a vacuum.

Special Theory of Relativity

By the late nineteenth century, various experimental anomalies could not be explained by the simple wave theory. One of these anomalies involved a controversy over the speed of light. The speed of light and other EMR predicted by Maxwell's equations did not appear unless the equations were modified in a way first suggested by FitzGerald and Lorentz, or else otherwise that speed would depend on the speed of observer relative to the "medium" (called luminiferous aether) which supposedly "carried" the electromagnetic wave (in a manner analogous to the way air carries sound waves). Experiments failed to find any observer effect. In 1905, Einstein proposed that space and time appeared to be velocity-changeable entities for light propagation and all other processes and laws. These changes accounted for the constancy of the speed of light and all electromagnetic radiation, from the viewpoints of all observers—even those in relative motion.

History of Discovery

Electromagnetic radiation of wavelengths other than those of visible light were discovered in the early 19th century. The discovery of infrared radiation is ascribed to astronomer William Herschel, who published his results in 1800 before the Royal Society of London. Herschel used a glass prism to refract light from the Sun and detected invisible rays that caused heating beyond the red part of the spectrum, through an increase in the temperature recorded with a thermometer. These "calorific rays" were later termed infrared.

In 1801, German physicist Johann Wilhelm Ritter discovered ultraviolet in an experiment similar to Hershel's, using sunlight and a glass prism. Ritter noted that invisible rays near the violet edge of a solar spectrum dispersed by a triangular prism darkened silver chloride preparations more quickly than did the nearby violet light. Ritter's experiments were an early precursor to what would become photography. Ritter noted that the ultraviolet rays (which at first were called "chemical rays") were capable of causing chemical reactions.

In 1862-4 James Clerk Maxwell developed equations for the electromagnetic field which suggested that waves in the field would travel with a speed that was very close to the known speed of light. Maxwell therefore suggested that visible light (as well as invisible infrared and ultraviolet rays by inference) all consisted of propagating disturbances (or radiation) in the electromagnetic field. Radio waves were first produced deliberately by Heinrich Hertz in 1887, using electrical circuits calculated to produce oscillations at a much lower frequency than that of visible light, following recipes for producing oscillating charges and currents suggested by Maxwell's equations. Hertz also developed ways to detect these waves, and produced and characterized what were later termed radio waves and microwaves.

Wilhelm Röntgen discovered and named X-rays. After experimenting with high voltages applied to an evacuated tube on 8 November 1895, he noticed a fluorescence on a nearby plate of coated glass. In one month, he discovered X-rays' main properties.

The last portion of the EM spectrum to be discovered was associated with radioactivity. Henri Becquerel found that uranium salts caused fogging of an unexposed photographic plate through a covering paper in a manner similar to X-rays, and Marie Curie discovered that only certain elements gave off these rays of energy, soon discovering the intense radiation of radium. The radiation from pitchblende was differentiated into alpha rays (alpha particles) and beta rays (beta particles) by

Ernest Rutherford through simple experimentation in 1899, but these proved to be charged particulate types of radiation. However, in 1900 the French scientist Paul Villard discovered a third neutrally charged and especially penetrating type of radiation from radium, and after he described it, Rutherford realized it must be yet a third type of radiation, which in 1903 Rutherford named gamma rays. In 1910 British physicist William Henry Bragg demonstrated that gamma rays are electromagnetic radiation, not particles, and in 1914 Rutherford and Edward Andrade measured their wavelengths, finding that they were similar to X-rays but with shorter wavelengths and higher frequency, although a 'cross-over' between X and gamma rays makes it possible to have X-rays with a higher energy (and hence shorter wavelength) than gamma rays and vice versa. The origin of the ray differentiates them, gamma rays tend to be a natural phenomena originating from the unstable nucleus of an atom and X-rays are electrically generated (and hence man-made) unless they are as a result of bremsstrahlung X-radiation caused by the interaction of fast moving particles (such as beta particles) colliding with certain materials, usually of higher atomic numbers.

Electromagnetic Spectrum

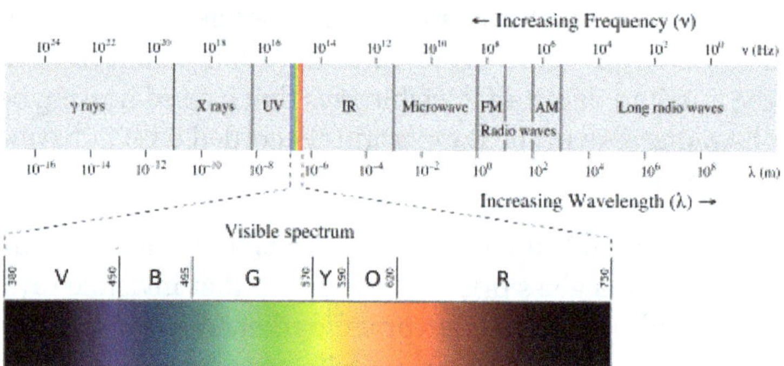

Electromagnetic spectrum with visible light highlighted

EM radiation (the designation 'radiation' excludes static electric and magnetic and near fields) is classified by wavelength into radio, microwave, infrared, visible, ultraviolet, X-rays and gamma rays. Arbitrary electromagnetic waves can be expressed by Fourier analysis in terms of sinusoidal monochromatic waves, which in turn can each be classified into these regions of the EMR spectrum.

For certain classes of EM waves, the waveform is most usefully treated as *random*, and then spectral analysis must be done by slightly different mathematical techniques appropriate to random or stochastic processes. In such cases, the individual frequency components are represented in terms of their *power* content, and the phase information is not preserved. Such a representation is called the power spectral density of the random process. Random electromagnetic radiation requiring this kind of analysis is, for example, encountered in the interior of stars, and in certain other very wideband forms of radiation such as the Zero point wave field of the electromagnetic vacuum.

The behavior of EM radiation depends on its frequency. Lower frequencies have longer wavelengths, and higher frequencies have shorter wavelengths, and are associated with photons of higher energy. There is no fundamental limit known to these wavelengths or energies, at either end of the spectrum, although photons with energies near the Planck energy or exceeding it (far too high to have ever been observed) will require new physical theories to describe.

Soundwaves are not electromagnetic radiation. At the lower end of the electromagnetic spectrum, about 20 Hz to about 20 kHz, are frequencies that might be considered in the audio range. However, electromagnetic waves cannot be directly perceived by human ears. Sound waves are instead the oscillating compression of molecules. To be heard, electromagnetic radiation must be converted to pressure waves of the fluid in which the ear is located (whether the fluid is air, water or something else).

Interactions as a Function of Frequency

When EM radiation interacts with matter, its behavior changes qualitatively as its frequency changes.

Radio and Microwave

At radio and microwave frequencies, EMR interacts with matter largely as a bulk collection of charges which are spread out over large numbers of affected atoms. In electrical conductors, such induced bulk movement of charges (electric currents) results in absorption of the EMR, or else separations of charges that cause generation of new EMR (effective reflection of the EMR). An example is absorption or emission of radio waves by antennas, or absorption of microwaves by water or other molecules with an electric dipole moment, as for example inside a microwave oven. These interactions produce either electric currents or heat, or both.

Infrared

Like radio and microwave, infrared also is reflected by metals (as is most EMR into the ultraviolet). However, unlike lower-frequency radio and microwave radiation, Infrared EMR commonly interacts with dipoles present in single molecules, which change as atoms vibrate at the ends of a single chemical bond. It is consequently absorbed by a wide range of substances, causing them to increase in temperature as the vibrations dissipate as heat. The same process, run in reverse, causes bulk substances to radiate in the infrared spontaneously.

Visible Light

As frequency increases into the visible range, photons have enough energy to change the bond structure of some individual molecules. It is not a coincidence that this happens in the "visible range," as the mechanism of vision involves the change in bonding of a single molecule (retinal) which absorbs light in the rhodopsin in the retina of the human eye. Photosynthesis becomes possible in this range as well, for similar reasons, as a single molecule of chlorophyll is excited by a single photon. Animals that detect infrared make use of small packets of water that change temperature, in an essentially thermal process that involves many photons. For this reason, infrared, microwaves and radio waves are thought to damage molecules and biological tissue only by bulk heating, not excitation from single photons of the radiation.

Visible light is able to affect a few molecules with single photons, but usually not in a permanent or damaging way, in the absence of power high enough to increase temperature to damaging levels. However, in plant tissues that conduct photosynthesis, carotenoids act to quench electronically excited chlorophyll produced by visible light in a process called non-photochemical quenching, in order to prevent reactions that would otherwise interfere with photosynthesis at high light levels. Limited evidence indicate that some reactive oxygen species are created by visible light in skin, and that these may have some role in photoaging, in the same manner as ultraviolet A.

Ultraviolet

As frequency increases into the ultraviolet, photons now carry enough energy (about three electron volts or more) to excite certain doubly bonded molecules into permanent chemical rearrangement. In DNA, this causes lasting damage. DNA is also indirectly damaged by reactive oxygen species produced by ultraviolet A (UVA), which has energy too low to damage DNA directly. This is why ultraviolet at all wavelengths can damage DNA, and is capable of causing cancer, and (for UVB) skin burns (sunburn) that are far worse than would be produced by simple heating (temperature increase) effects. This property of causing molecular damage that is out of proportion to heating effects, is characteristic of all EMR with frequencies at the visible light range and above. These properties of high-frequency EMR are due to quantum effects that permanently damage materials and tissues at the molecular level.

At the higher end of the ultraviolet range, the energy of photons becomes large enough to impart enough energy to electrons to cause them to be liberated from the atom, in a process called photoionisation. The energy required for this is always larger than about 10 electron volts (eV) corresponding with wavelengths smaller than 124 nm (some sources suggest a more realistic cutoff of 33 eV, which is the energy required to ionize water). This high end of the ultraviolet spectrum with energies in the approximate ionization range, is sometimes called "extreme UV." Ionizing UV is strongly filtered by the Earth's atmosphere).

X-rays and Gamma Rays

Electromagnetic radiation composed of photons that carry minimum-ionization energy, or more, (which includes the entire spectrum with shorter wavelengths), is therefore termed ionizing radiation. (Many other kinds of ionizing radiation are made of non-EM particles). Electromagnetic-type ionizing radiation extends from the extreme ultraviolet to all higher frequencies and shorter wavelengths, which means that all X-rays and gamma rays qualify. These are capable of the most severe types of molecular damage, which can happen in biology to any type of biomolecule, including mutation and cancer, and often at great depths below the skin, since the higher end of the X-ray spectrum, and all of the gamma ray spectrum, penetrate matter.

Atmosphere and Magnetosphere

Most UV and X-rays are blocked by absorption first from molecular nitrogen, and then (for wavelengths in the upper UV) from the electronic excitation of dioxygen and finally ozone at the mid-range of UV. Only 30% of the Sun's ultraviolet light reaches the ground, and almost all of this is well transmitted.

Visible light is well transmitted in air, as it is not energetic enough to excite nitrogen, oxygen, or ozone, but too energetic to excite molecular vibrational frequencies of water vapor.

Absorption bands in the infrared are due to modes of vibrational excitation in water vapor.

However, at energies too low to excite water vapor, the atmosphere becomes transparent again, allowing free transmission of most microwave and radio waves.

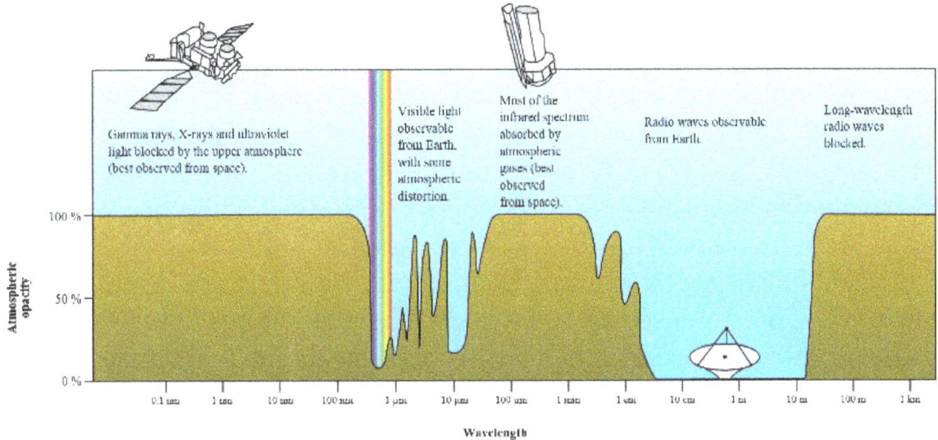

Rough plot of Earth's atmospheric absorption and scattering (or opacity) of various wavelengths of electromagnetic radiation

Finally, at radio wavelengths longer than 10 meters or so (about 30 MHz), the air in the lower atmosphere remains transparent to radio, but plasma in certain layers of the ionosphere begins to interact with radio waves. This property allows some longer wavelengths (100 meters or 3 MHz) to be reflected and results in shortwave radio beyond line-of-sight. However, certain ionospheric effects begin to block incoming radiowaves from space, when their frequency is less than about 10 MHz (wavelength longer than about 30 meters).

Types and Sources, Classed by Spectral Band

Radio Waves

Radio waves have the least amount of energy and the lowest frequency. When radio waves impinge upon a conductor, they couple to the conductor, travel along it and induce an electric current on the conductor surface by moving the electrons of the conducting material in correlated bunches of charge. Such effects can cover macroscopic distances in conductors (such as radio antennas), since the wavelength of radiowaves is long.

Microwaves

Microwaves are a form of electromagnetic radiation with wavelengths ranging from as long as one meter to as short as one millimeter; with frequencies between 300 MHz (0.3 GHz) and 300 GHz.

Visible Light

Natural sources produce EM radiation across the spectrum. EM radiation with a wavelength between approximately 400 nm and 700 nm is directly detected by the human eye and perceived as visible light. Other wavelengths, especially nearby infrared (longer than 700 nm) and ultraviolet (shorter than 400 nm) are also sometimes referred to as light.

Thermal Radiation and Electromagnetic Radiation as a form of Heat

The basic structure of matter involves charged particles bound together. When electromagnetic radiation impinges on matter, it causes the charged particles to oscillate and gain energy. The ultimate fate of this energy depends on the context. It could be immediately re-radiated and appear as scattered, reflected, or transmitted radiation. It may get dissipated into other microscopic motions within the matter, coming to thermal equilibrium and manifesting itself as thermal energy in the material. With a few exceptions related to high-energy photons (such as fluorescence, harmonic generation, photochemical reactions, the photovoltaic effect for ionizing radiations at far ultraviolet, X-ray and gamma radiation), absorbed electromagnetic radiation simply deposits its energy by heating the material. This happens for infrared, microwave and radio wave radiation. Intense radio waves can thermally burn living tissue and can cook food. In addition to infrared lasers, sufficiently intense visible and ultraviolet lasers can easily set paper afire.

Ionizing radiation creates high-speed electrons in a material and breaks chemical bonds, but after these electrons collide many times with other atoms eventually most of the energy becomes thermal energy all in a tiny fraction of a second. This process makes ionizing radiation far more dangerous per unit of energy than non-ionizing radiation. This caveat also applies to UV, even though almost all of it is not ionizing, because UV can damage molecules due to electronic excitation, which is far greater per unit energy than heating effects.

Infrared radiation in the spectral distribution of a black body is usually considered a form of heat, since it has an equivalent temperature and is associated with an entropy change per unit of thermal energy. However, "heat" is a technical term in physics and thermodynamics and is often confused with thermal energy. Any type of electromagnetic energy can be transformed into thermal energy in interaction with matter. Thus, *any* electromagnetic radiation can "heat" (in the sense of increase the thermal energy termperature of) a material, when it is absorbed.

The inverse or time-reversed process of absorption is thermal radiation. Much of the thermal energy in matter consists of random motion of charged particles, and this energy can be radiated away from the matter. The resulting radiation may subsequently be absorbed by another piece of matter, with the deposited energy heating the material.

The electromagnetic radiation in an opaque cavity at thermal equilibrium is effectively a form of thermal energy, having maximum radiation entropy.

Biological Effects

Bioelectromagnetics is the study of the interactions and effects of EM radiation on living organisms. The effects of electromagnetic radiation upon living cells, including those in humans, depends upon the radiation's power and frequency. For low-frequency radiation (radio waves to visible light) the best-understood effects are those due to radiation power alone, acting through heating when radiation is absorbed. For these thermal effects, frequency is important only as it affects penetration into the organism (for example, microwaves penetrate better than infrared). Initially, it was believed that low frequency fields that were too weak to cause significant heating could not possibly have any biological effect.

Despite this opinion among researchers, evidence has accumulated that supports the existence of complex biological effects of weaker *non-thermal* electromagnetic fields, (including weak ELF magnetic fields, although the latter does not strictly qualify as EM radiation), and modulated RF and microwave fields. Fundamental mechanisms of the interaction between biological material and electromagnetic fields at non-thermal levels are not fully understood.

The World Health Organization has classified radio frequency electromagnetic radiation as Group 2B - possibly carcinogenic. This group contains possible carcinogens that have weaker evidence, at the same level as coffee and automobile exhaust. For example, epidemiological studies looking for a relationship between cell phone use and brain cancer development, have been largely inconclusive, save to demonstrate that the effect, if it exists, cannot be a large one.

At higher frequencies (visible and beyond), the effects of individual photons begin to become important, as these now have enough energy individually to directly or indirectly damage biological molecules. All UV frequences have been classed as Group 1 carcinogens by the World Health Organization. Ultraviolet radiation from sun exposure is the primary cause of skin cancer.

Thus, at UV frequencies and higher (and probably somewhat also in the visible range), electromagnetic radiation does more damage to biological systems than simple heating predicts. This is most obvious in the "far" (or "extreme") ultraviolet. UV, with X-ray and gamma radiation, are referred to as ionizing radiation due to the ability of photons of this radiation to produce ions and free radicals in materials (including living tissue). Since such radiation can severely damage life at energy levels that produce little heating, it is considered far more dangerous (in terms of damage-produced per unit of energy, or power) than the rest of the electromagnetic spectrum.

Derivation from Electromagnetic Theory

Electromagnetic waves were predicted by the classical laws of electricity and magnetism, known as Maxwell's equations. Inspection of Maxwell's equations without sources (charges or currents) results in nontrivial solutions of changing electric and magnetic fields. Beginning with Maxwell's equations in free space:

$$\nabla \cdot \mathbf{E} = 0$$

$$\nabla \times \mathbf{E} = -\frac{\partial \mathbf{B}}{\partial t}$$

$$\nabla \cdot \mathbf{B} = 0$$

$$\nabla \times \mathbf{B} = \mu_0 \epsilon_0 \frac{\partial \mathbf{E}}{\partial t}$$

where

∇ is a vector differential operator.

One solution,

$$\mathbf{E}=\mathbf{B}=\mathbf{0},$$

is trivial.

For a more useful solution, we utilize vector identities, which work for any vector, as follows:

$$\nabla\times(\nabla\times\mathbf{A})=\nabla(\nabla\cdot\mathbf{A})-\nabla^2\mathbf{A}$$

The curl of equation:

$$\nabla\times(\nabla\times\quad)=\nabla\times\left(-\frac{\partial}{\partial}\right)$$

Evaluating the left hand side:

$$\nabla\times(\nabla\times\mathbf{E})=\nabla(\nabla\cdot\mathbf{E})-\nabla^2\mathbf{E}=-\nabla^2\mathbf{E}$$

simplifying the above by using equation.

Evaluating the right hand side:

$$\nabla\times\left(-\frac{\partial\mathbf{B}}{\partial t}\right)=-\frac{\partial}{\partial t}(\nabla\times\mathbf{B})=-\mu_0\epsilon_0\frac{\partial^2\mathbf{E}}{\partial t^2}$$

Equations are equal, so this results in a vector-valued differential equation for the electric field, namely

$$\nabla^2\mathbf{E}=\mu_0\epsilon_0\frac{\partial^2\mathbf{E}}{\partial t^2}$$

Applying a similar pattern results in similar differential equation for the magnetic field:

$$\nabla^2\mathbf{B}=\mu_0\epsilon_0\frac{\partial^2\mathbf{B}}{\partial t^2}.$$

These differential equations are equivalent to the wave equation:

$$\nabla^2 f=\frac{1}{c_0^2}\frac{\partial^2 f}{\partial t^2}$$

where

c_o is the speed of the wave in free space and

f describes a displacement

Or more simply:

$$\Box f = 0$$

where $\Box$ is d'Alembertian:

$$\Box = \nabla^2 - \frac{1}{c_0^2}\frac{\partial^2}{\partial t^2} = \frac{\partial^2}{\partial x^2} + \frac{\partial^2}{\partial y^2} + \frac{\partial^2}{\partial z^2} - \frac{1}{c_0^2}\frac{\partial^2}{\partial t^2}$$

In the case of the electric and magnetic fields, the speed is:

$$c_0 = \frac{1}{\sqrt{\mu_0 \epsilon_0}}$$

This is the speed of light in vacuum. Maxwell's equations unified the vacuum permittivity ϵ_0, the vacuum permeability μ_0, and the speed of light itself, c_0. This relationship had been discovered by Wilhelm Eduard Weber and Rudolf Kohlrausch prior to the development of Maxwell's electrodynamics, however Maxwell was the first to produce a field theory consistent with waves traveling at the speed of light.

These are only two equations versus the original four, so more information pertains to these waves hidden within Maxwell's equations. A generic vector wave for the electric field.

$$\mathbf{E} = \mathbf{E}_0 f\left(\hat{\mathbf{k}} \cdot \mathbf{x} - c_0 t\right)$$

Here, $\mathbf{E}_0$ is the constant amplitude, f is any second differentiable function, $\hat{\mathbf{k}}$ is a unit vector in the direction of propagation, and $\mathbf{x}$ is a position vector. $f\left(\hat{\mathbf{k}} \cdot \mathbf{x} - c_0 t\right)$ is a generic solution to the wave equation. In other words,

$$\nabla^2 f\left(\hat{\mathbf{k}} \cdot \mathbf{x} - c_0 t\right) = \frac{1}{c_0^2}\frac{\partial^2}{\partial t^2} f\left(\hat{\mathbf{k}} \cdot \mathbf{x} - c_0 t\right),$$

for a generic wave traveling in the $\hat{\mathbf{k}}$ direction.

This form will satisfy the wave equation.

$$\nabla \cdot \mathbf{E} = \hat{\mathbf{k}} \cdot \mathbf{E}_0 f'\left(\hat{\mathbf{k}} \cdot \mathbf{x} - c_0 t\right) = 0$$

$$\mathbf{E} \cdot \hat{\mathbf{k}} = 0$$

The first of Maxwell's equations implies that the electric field is orthogonal to the direction the wave propagates.

$$\nabla \times \mathbf{E} = \hat{\mathbf{k}} \times \mathbf{E}_0 f'\left(\hat{\mathbf{k}} \cdot \mathbf{x} - c_0 t\right) = -\frac{\partial \mathbf{B}}{\partial t}$$

$$\mathbf{B} = \frac{1}{c_0}\hat{\mathbf{k}} \times \mathbf{E}$$

The second of Maxwell's equations yields the magnetic field. The remaining equations will be satisfied by this choice of $\mathbf{E},\mathbf{B}$.

The electric and magnetic field waves in the far-field travel at the speed of light. They have a special restricted orientation and proportional magnitudes, $E_0 = c_0 B_0$, which can be seen immediately from the Poynting vector. The electric field, magnetic field, and direction of wave propagation are all orthogonal, and the wave propagates in the same direction as $\mathbf{E}\times\mathbf{B}$. Also, E and B far-fields in free space, which as wave solutions depend primarily on these two Maxwell equations, are in-phase with each other. This is guaranteed since the generic wave solution is first order in both space and time, and the curl operator on one side of these equations results in first-order spatial derivatives of the wave solution, while the time-derivative on the other side of the equations, which gives the other field, is first order in time, resulting in the same phase shift for both fields in each mathematical operation.

From the viewpoint of an electromagnetic wave traveling forward, the electric field might be oscillating up and down, while the magnetic field oscillates right and left. This picture can be rotated with the electric field oscillating right and left and the magnetic field oscillating down and up. This is a different solution that is traveling in the same direction. This arbitrariness in the orientation with respect to propagation direction is known as polarization. On a quantum level, it is described as photon polarization. The direction of the polarization is defined as the direction of the electric field.

More general forms of the second-order wave equations given above are available, allowing for both non-vacuum propagation media and sources. Many competing derivations exist, all with varying levels of approximation and intended applications. One very general example is a form of the electric field equation, which was factorized into a pair of explicitly directional wave equations, and then efficiently reduced into a single uni-directional wave equation by means of a simple slow-evolution approximation

Electromagnetic Waves

What is light, particle or wave? Much of our daily experience with light, particularly the fact that light rays move in straight lines tells us that we can think of light as a stream of particles. This is further borne out when place an opaque object in the path of the light rays. The shadow, as shown in Figure is a projected image of the object, which is what we expect if light were a stream of partiles. But a closer look at the edges of the shadow reveals a very fine pattern of dark and bright bands or fringes. Such a pattern can also be seen if we stretch out our hand and look at the sky through a thin gap produced by bringing two of our fingers close. This cannot be explained unless we accept that light is some kind of a wave.

It is now well known that light is an electromagnetic wave. We shall next discuss what we mean by an electromagnetic wave or radiation.

Electromagnetic Radiation

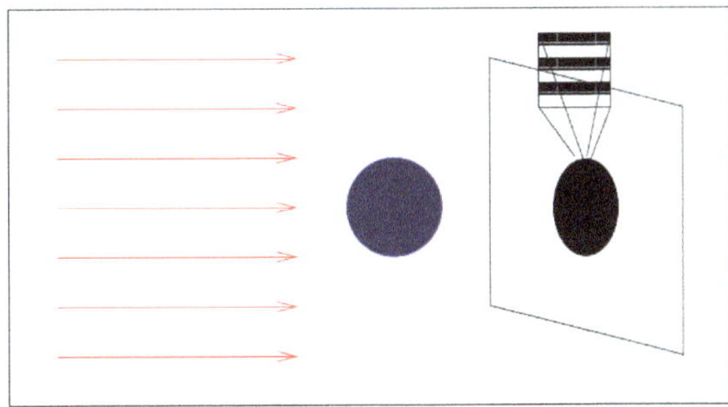

Shadow of an opaque object.

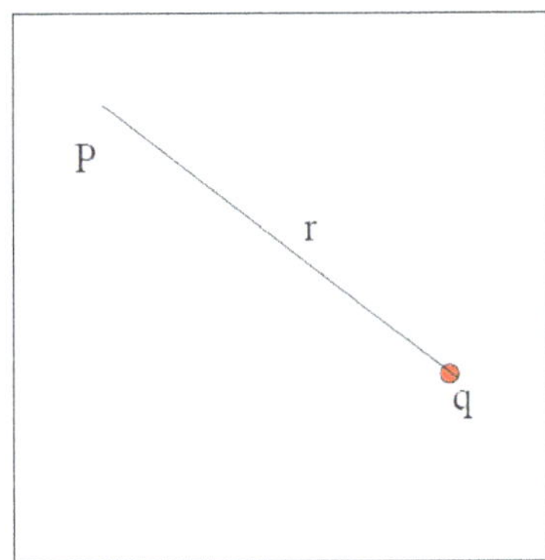

A charge placed at a distance r from the point of observation P.

What is the electric field produced at a point P by a charge q located at a distance r as shown in Figure? Anybody with a little knowledge of physics will tell us that this is given by Coulomb's law

$$\vec{E} = \frac{-q}{4\pi \in_0} \frac{\hat{e}_r}{r^2}$$

where $\hat{e}_r$ is an unit vector from P to the position of the charge. In what follows we shall follow the notation used in Feynman Section.

In the 1880s J.C. Maxwell proposed a modification in the laws of electricity and magnetism which were known at that time. The change proposed by Maxwell unified our ideas of electricity and magnetism and showed both of them to be manifestations of a single underlying quantity. Further it implied that Coulomb's law did not tells us the complete picture. The correct formula for the electric field is

$$\vec{E} = \frac{-q}{4\pi \in_0} \left[\frac{\hat{e}_{r'}}{r'^2} + \frac{r'}{c} \frac{d}{dt} \left(\frac{\hat{e}_{r'}}{r'^2} \right) + \frac{1}{c^2} \frac{d^2}{dt^2} \hat{e}_{r'} \right]$$

This formula incorporates several new effects. The first is the fact that no information can propagate instantaneously. This is a drawback of Coulomb's law where the electric field at a distant point P changes the moment the position of the charge is changed. This should actually happen after some time. The new formula incorporates the fact that the influence of the charge propagates at a speed c. The electric field at the time t is determined by the position of the charge at an earlier time. This is referred to as the retarded position of the charge r', and $\hat{e}_{r'}$ also refers to the retarded position.

The first term in eq. is Coulomb's law with the retarded position. In addition there are two new terms which arise due to the modification proposed by Maxwell. These two terms contribute only when the charge moves. The magnetic field produced by the charge is

$$\vec{B} = -\hat{e}_{r'} \times \vec{E} / c$$

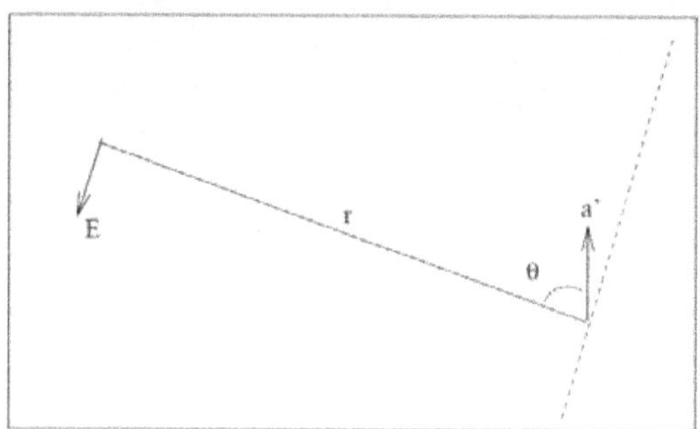

Electric field for an accelerated charge.

A close look at eq. shows that the contribution from the first two terms falls off as $1/r'^2$ and these two terms are of not of interest at large distances from the charge. It is only the third term which has a $1/r$ behaviour that makes a significant contribution at large distances. This term permits a charged particle to influence another charged particle at a great distance through the $1/r$ electric field. This is referred to as electromagnetic radiation and light is a familiar example of this phenomenon. It is obvious from the formula that only accelerating charges produce radiation.

The interpretation of the formula is substantially simplified if we assume that the motion of the charge is relatively slow, and is restricted to a region which is small in comparison to the distance r to the point where we wish to calculate the electric field. We then have

$$\frac{d^2}{dt^2}\hat{e}_{r'} = \frac{d^2}{dt^2}\left(\frac{\vec{r}'}{r'}\right) \approx \frac{\ddot{\vec{r}}'_{\perp}}{r}$$

where $\ddot{\vec{r}}'_{\perp}$ is the acceleration of the charge in the direction perpendicular to $\hat{e}_{r'}$. The parallel component of the acceleration does not effect the unit vector $\hat{e}_{r'}$. and hence it does not make a contribution here. Further, the motion of the charge makes a very small contribution to r' in the denominator, neglecting this we replace r' with the constant distance r.

The electric field at a time t is related to $a(t-r/c)$ which is the retarded acceleration as

$$E(t) = \frac{-q}{4\pi \epsilon_0 c^2 r} a(t - r/c) \sin\theta$$

Where θ is the angle between the line of sight $\hat{e}_r$ to the charge and the direction of the retarded acceleration vector. The electric field vector is in the direction obtained by projecting the retarded acceleration vector on the plane perpendicular to $\hat{e}_r$ as shown in Figure.

Problem 1: Show that the second term inside the bracket of eq. indeed falls off as $1/r'^2$. Also show that the expression for electric field for an accelerated charge i.e. eq. follows from it.

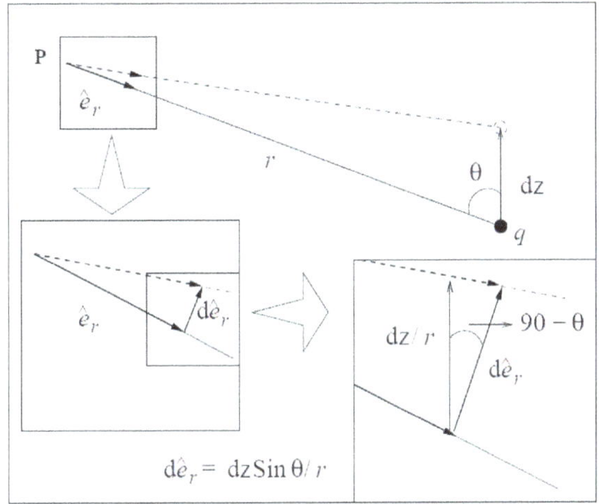

Solution to Problem 1.

Solution 1: See figure (r and θ can be treated as constants with respect to time).

Electric Dipole Radiation

We next consider a situation where a charge accelerates up and down along a straight line. The analysis of this situation using equation has wide applications including many in technology. We consider the device shown in Figure which has two wires A and B connected to an oscillating voltage generator. Consider the situation when the terminal of the voltage generator connected

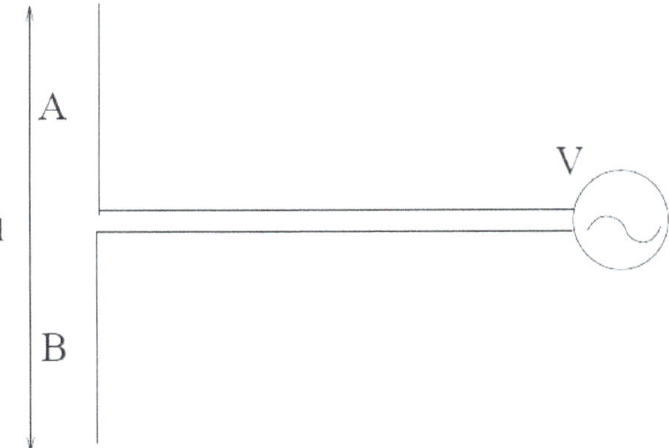

An electric dipole.

to A is positive and the one connected to B is negative. There will be an accumulation of positive charge at the tip of the wire A and negative charge at the tip of B respectively. The electrons rush from B to A when the voltage is reversed. The oscillating voltage causes charge to oscillate up and down the two wire A and B as if they were a single wire. In the situation where the time taken by the electrons to move up and down the wires is much larger than the time taken for light signal to cross the wire, this can be thought of as an oscillating electric dipole. Note that here we have many electrons oscillating up and down the wire. Since all the electrons have the same acceleration, the electric fields that they produce adds up.

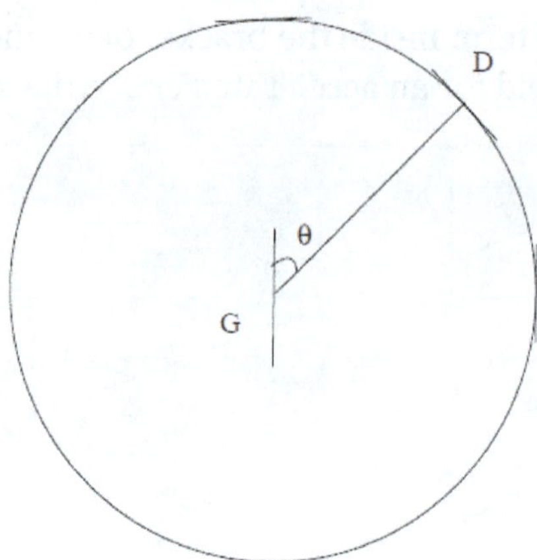

A dipole with a detection.

The electric field produced is inversely proportional to the distance r from the oscillator. At any time t, the electric field is proportional to the acceleration of the charges at a time $t - r/c$ in the past.

It is possible to measure the radiation using a another electric dipole oscillator where the voltage generator is replaced by a detector, say an oscilloscope. An applied oscillating electric field will give rise to an oscillating current in the wires which can be converted to a voltage and measured. A dipole can measure oscillating electric fields only if the field is parallel to the dipole and not if they are perpendicular. A dipole is quite commonly used as an antenna to receive radio waves which is a form of electromagnetic radiation.

Figure shows an experiment where we use a dipole with a detector (D) to measure the electromagnetic radiation produced by another oscillating electric dipole (G). The detected voltage is maximum at $\theta = 90°$ and falls as sin θ in other directions. At any point on the circle, the direction of the electric field vector of the emitted radiation is along the tangent. Often we find that placing the antenna of a transistor radio in a particular orientation improves the reception. This is roughly aligning the antenna with the incoming radiation which was transmitted by a transmitter.

Sinusoidal Oscillations

We next consider a situation where a sinusoidal voltage is applied to the dipole oscillator. The dipole is aligned with the y axis. The voltage causes the charges to move up and down as

$$y(t) = y_0 \cos(\omega t),$$

with acceleration

$$a(t) = -\omega^2 y_0 \cos(\omega t),$$

producing an electric field

$$E(t) = \frac{q y_0 \omega^2}{4\pi \epsilon_0 c^2 r} \cos\left[\omega(t - r/c)\right] \sin\theta.$$

It is often useful and interesting to represent the oscillating charge in terms of other equivalent quantities namely the dipole moment and the current in the circuit. Let us replace the charge q which moves up and down as $y(t)$ by two charges, one charge $q/2$ which moves as $y(t)$ and another charge $-q/2$ which moves in exactly the opposite direction as $-y(t)$ .The electric field produced by the new configuration is exactly the same as that produced by the single charge considered earlier. This allows us to interpret equation in terms of an oscillating dipole

$$d_y(t) = q\,y(t) = d_0 \cos(\omega t),$$

which allows us to write equation as

$$E(t) = \frac{-1}{4\pi \in_0 c^2 r} \ddot{d}_y(t - r/c)\sin\theta.$$

Returning once more to the dipole oscillator shown in Figure, we note that the excess electrons which rush from A to B when B has a positive voltage reside at the tip of the wire B. Further, there is an equal excess of positive charge in A which resides at the tip of A. The fact that excess charge resides at the tips of the wire is property of charges on conductors which should be familiar from the study of electrostatics. Now the dipole moment is $d_y(t) = l\,q(t)$ where l is the length of the dipole oscillator and $q(t)$ is the excess charge accumulated at one of the tips. This allows us to write $\ddot{d}_y(y)$ in terms of the current in the wires $I(t) = \dot{q}(t)$ as

$$\ddot{d}_y(t) = l\dot{I}(t).$$

We can then express the electric field produced by the dipole oscillator in terms of the current. This is particularly useful when considering technological applications of the electric dipole oscillator. For a current

$$I(t) = -I\sin(\omega t)$$

the electric field is given by

$$E(t) = \frac{\omega l I}{4\pi \in_0 c^2 r} \cos\left[\omega(t - r/c)\right]\sin\theta.$$

where I refers to the peak current in the wire.

We now end the small detour where we discussed how the electric field is related to the dipole moment and the current, and return to our discussion of the electric field predicted by equation. We shall restrict our attention to points along the x axis. The electric field of the radiation is in the y direction and has a value (by substituting $r = ax$ and $\theta = \pi/2$)

$$E_y(x,t) = \frac{q y_0 \omega^2}{4\pi \in_0 c^2 x} \cos\left[\omega t - \left(\frac{\omega}{c}\right)x\right]$$

Let us consider a situation where we are interested in the x dependence of the electric field at a great distance from the emitter. Say we are $1\,km$ away from the oscillator and we would like to know how the electric field varies at two points which are $1m$ apart. This situation is shown schematically in Figure. The point to note is that a small variation in x will make a very small difference to the $1/x$ dependence of the electric field which we can neglect, but the change in the cos term cannot be neglected. This is because x is multiplied by a factor ω/c which could be large and a change in $\omega x/c$ would mean a different phase of the oscillation. Thus at large distances the electric field of the radiation can be well described by

$$E_y(x,t) = E\cos[\omega t - kx]$$

where the wave number is $k = \omega/c$. This is the familiar sinusoidal plane wave which we have studied in an earlier chapter and which can be represented in the complex notation as

$$\tilde{E}_y(x,t) = \tilde{E}e^{i[\omega t - kx]}$$

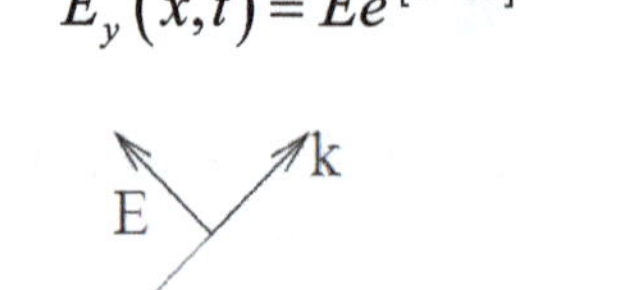

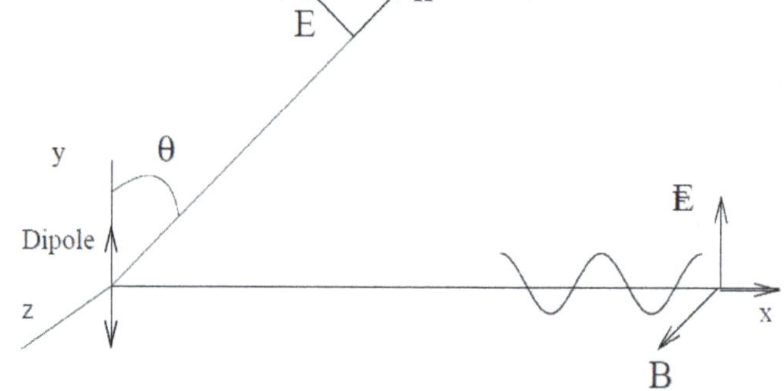

A dipole along the y axis

We next calculate the magnetic field $\vec{B}$. Referring to Figure we see that we have $\hat{e}_r = -\hat{i}$. Using this in equation with $\vec{E} = E_y(x,t)\hat{j}$ we have

$$\vec{B}(x,t) = \hat{i} \times \vec{E}_y(x,t)/c = \frac{E}{c}\cos(\omega t - kx)\hat{k}$$

The magnetic field is perpendicular to $\vec{E}(x,t)$ and its amplitude is a factor $1/c$ smaller than the electric field. The magnetic field oscillates with the same frequency and phase as the electric field.

Although our previous discussion was restricted to points along the x axis, the facts which we have learnt about the electric and magnetic fields hold at any position. At any point the direction of the electromagnetic wave is radially outwards with wave vector $\hat{k} = k\hat{r}$. The electric and magnetic fields are mutually perpendicular, they are also perpendicular to the wave vector $\hat{k}$.

Energy Density, Flux and Power

We now turn our attention to the energy carried by the electromagnetic wave. For simplicity we shall initially restrict ourselves to points located along the x axis for the situation shown in Figure.

The energy density in the electric and magnetic fields is given by

$$U = \frac{1}{2} \epsilon_0 E^2 + \frac{1}{2\mu_0} B^2$$

For an electromagnetic wave the electric and magnetic fields are related. The energy density can then be written in terms of only the electric field as

$$U = \frac{1}{2} \epsilon_0 E^2 + \frac{1}{2c^2 \mu_0} E^2$$

The speed of light c is related to ϵ_0 and μ_0 as $c^2 = 1/\epsilon_0 \mu_0$. Using this we find that the energy density has the form

$$U = \epsilon_0 E^2$$

The instantaneous energy density of the electromagnetic wave oscillates with time. The time average of the energy density is often a more useful quantity. We have already discussed how to calculate the time average of an oscillating quantity. This is particularly simple in the complex notation where the electric field

$$\tilde{E}_y(x,t) = \tilde{E} e^{i(\omega t - kx)}$$

has a mean squared value $\langle E^2 \rangle = \tilde{E}\tilde{E}^* / 2$. Using this we find that the average energy density is

$$\langle U \rangle = \frac{1}{2} \epsilon_0 \tilde{E}\tilde{E}^* = \frac{1}{2} \epsilon_0 E^2$$

where E is the amplitude of the electric field.

We next consider the energy flux of the electromagnetic radiation. The radiation propagates along the x axis at the point where we want to calculate the flux. Consider a surface which is perpendicular to direction in which the wave is propagating as shown in Figure. The energy flux S refers to the energy which crosses an unit area of this surface in unit time. It has units $Watt\, m^{-2}$. The flux S is the power that would be received by collecting the radiation in an area $1 m^2$ placed perpendicular to the direction in which the wave is propagating as shown in Figure.

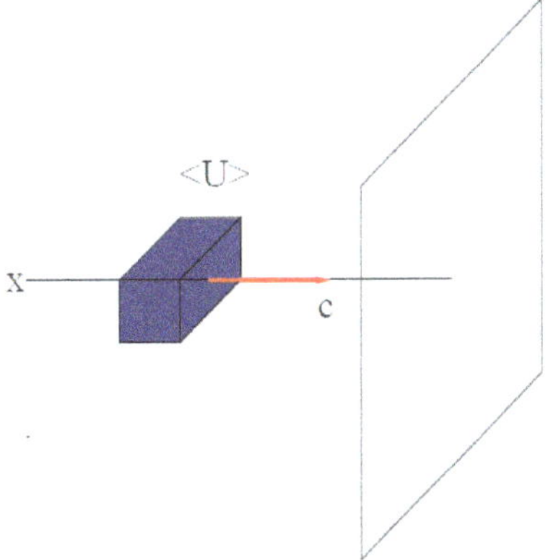

Flux through unit area perpendicular to the direction of propagation.

The average flux can be calculated by noting that the wave propagates along the x axis with speed c. The average energy $\langle U \rangle$ contained in an unit volume would take a time $1/c$ to cross the surface. The flux is the energy which would cross in one second which is

$$\langle S \rangle = \langle U \rangle c = \frac{1}{2} c \in_0 E^2$$

The energy flux is actually a vector quantity representing both the direction and the rate at which the wave carries energy. Referring back to equations we see that at any point the average flux $\langle \vec{S} \rangle$ is pointed

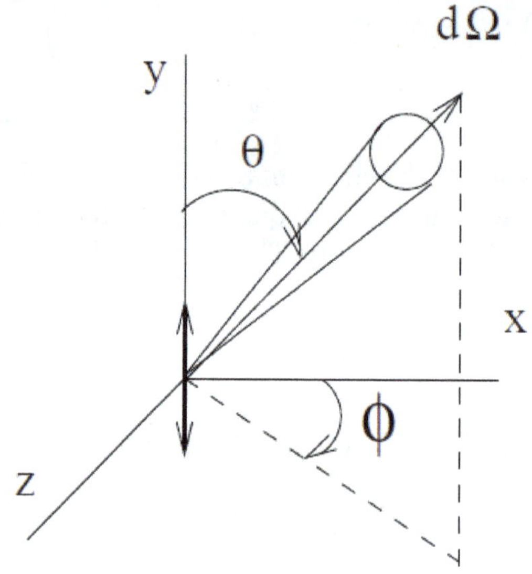

Solid angle d Ω.

radially outwards and has a value

$$\langle \vec{S} \rangle = \left(\frac{q^2 y_0^2 \omega^4}{32\pi^2 c^3 \in_0} \right) \frac{\sin^2 \theta}{r^2} \hat{r}.$$

Note that the flux falls as $1/r^2$ as we move away from the source. This is a property which may already be familiar to some of us from considerations of the conservation of energy. Note that the total energy crossing a surface enclosing the source will be constant irrespective of the shape and size of the surface.

Let us know shift our point of reference to the location of the dipole and ask how much power is radiated in any given direction. This is quantified using the power emitted per solid angle. Consider a solid angle $d\Omega$ along a direction $\hat{r}$ at an angle θ to the dipole as shown in Figure. The power dP radiated into this solid angle can be calculated by multiplying the flux with the area corresponding to this solid angle

$$dP = \langle \vec{S} \rangle \cdot \hat{r} r^2 d\Omega$$

which gives us the power radiated per unit solid angle to be

$$\frac{d}{d\Omega}\langle\vec{P}\rangle(\theta)=\left(\frac{q^2 y_0^2\omega^4}{32\pi^2c^3\in_0}\right)\sin^2\theta.$$

This tells us the radiation pattern of the dipole radiation, ie.. the directional dependence of the radiation is proportional to $\sin^2\theta$ .The radiation is maximum in the direction perpendicular to the dipole while there is no radiation emitted along the direction of the dipole. The radiation pattern is shown in Figure. Another important point to note is that the radiation depends on ω^4 which tells us that the same dipole will radiate significantly more power if it is made to oscillate at a higher frequency, doubling the frequency will increase the power sixteen times.

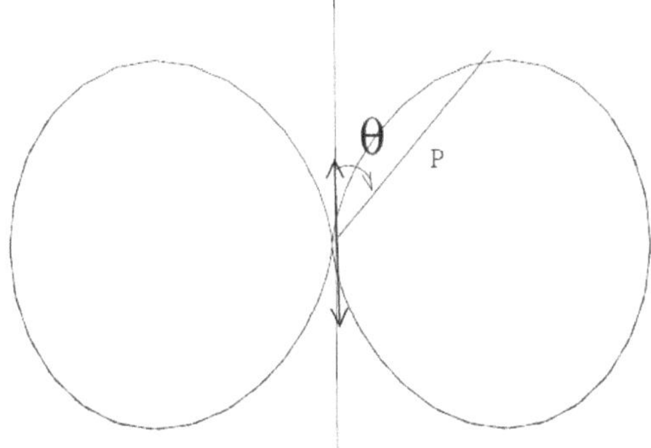

Dipole radiation pattern.

The total power radiated can be calculated by integrating over all solid angles. Using $d\Omega = sin\theta\ d\theta\ d\phi$ and

$$\int_0^{2\pi} d\phi\int_0^{\pi}\sin^3\theta\ d\theta=\frac{8\pi}{3}$$

gives the total power P to be

$$\langle\vec{P}\rangle=\frac{q^2y_0^2\omega^4}{12\pi c^3\in_0}.$$

It is often convenient to express this in terms of the amplitude of the current in the wires of the oscillator as

$$\langle\vec{P}\rangle=\frac{I^2l^2\omega^2}{12\pi c^3\in_0}.$$

The power radiated by the electric dipole is proportional to the square of the current. This behaviour is exactly the same as that of a resistance except that the oscillator emits the power as radiation while the resistance converts it to heat. We can express the radiated power in terms of an equivalent resistance with

$$\langle\vec{P}\rangle=\frac{1}{2}RI^2$$

Where

$$R = \left(\frac{l}{\lambda}\right)^2 790\Omega$$

l being the length of the dipole and λ the wavelength of the radiation.

Dipole

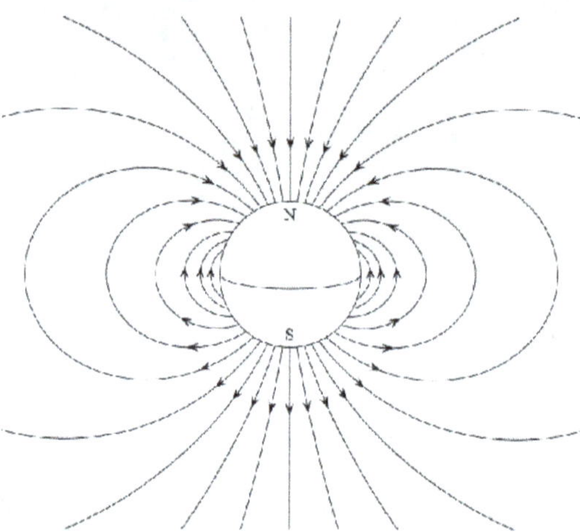

The magnetic field of a sphere with a south magnetic pole at the top and a north magnetic pole at the bottom. By comparison, Earth has a *south* magnetic pole near its north geographic pole and a *north* magnetic pole near its south pole.

In electromagnetism, there are two kinds of dipoles:

- An electric dipole is a separation of positive and negative charges. The simplest example of this is a pair of electric charges of equal magnitude but opposite sign, separated by some (usually small) distance. A permanent electric dipole is called an electret.
- A magnetic dipole is a closed circulation of electric current. A simple example of this is a single loop of wire with some constant current through it.

Dipoles can be characterized by their dipole moment, a vector quantity. For the simple electric dipole given above, the electric dipole moment points from the negative charge towards the positive charge, and has a magnitude equal to the strength of each charge times the separation between the charges. (To be precise: for the definition of the dipole moment, one should always consider the "dipole limit", where, for example, the distance of the generating charges should *converge* to 0 while simultaneously, the charge strength should *diverge* to infinity in such a way that the product remains a positive constant.)

For the current loop, the magnetic dipole moment points through the loop (according to the right hand grip rule), with a magnitude equal to the current in the loop times the area of the loop.

In addition to current loops, the electron, among other fundamental particles, has a magnetic dipole moment. That is because it generates a magnetic field that is identical to that generated by a very small current loop. However, the electron's magnetic moment is not due to a current loop, but is instead an intrinsic property of the electron. It is also possible that the electron has an *electric* dipole moment although it has not yet been observed.

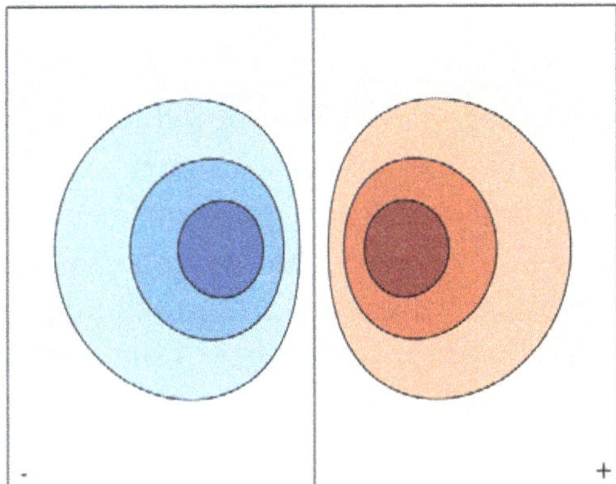

Contour plot of the electrostatic potential of a horizontally oriented electrical dipole of finite size. Strong colors indicate highest and lowest potential (where the opposing charges of the dipole are located).

A permanent magnet, such as a bar magnet, owes its magnetism to the intrinsic magnetic dipole moment of the electron. The two ends of a bar magnet are referred to as poles, and may be labeled "north" and "south". In terms of the Earth's magnetic field, they are respectively "north-seeking" and "south-seeking" poles: if the magnet were freely suspended in the Earth's magnetic field, the north-seeking pole would point towards the north and the south-seeking pole would point towards the south. The dipole moment of the bar magnet points from its magnetic south to its magnetic north pole. The north pole of a bar magnet in a compass points north. However, that means that Earth's geomagnetic north pole is the *south* pole (south-seeking pole) of its dipole moment and vice versa.

The only known mechanisms for the creation of magnetic dipoles are by current loops or quantum-mechanical spin since the existence of magnetic monopoles has never been experimentally demonstrated.

Classification

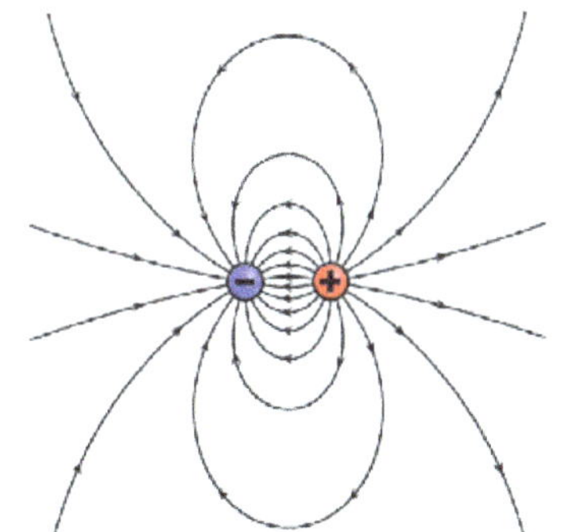

Electric field lines of two opposing charges separated by a finite distance.

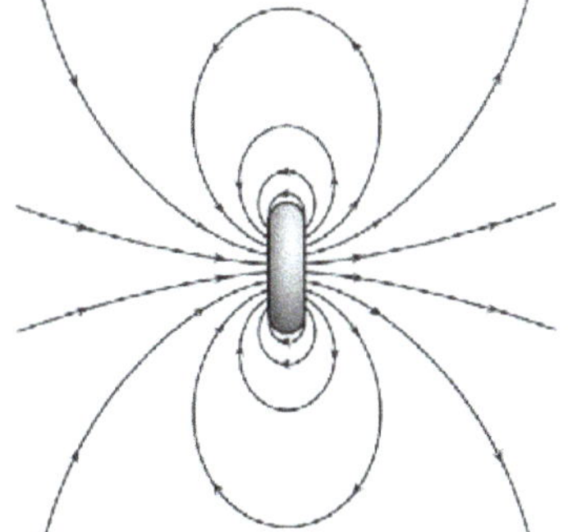

Magnetic field lines of a ring current of finite diameter.

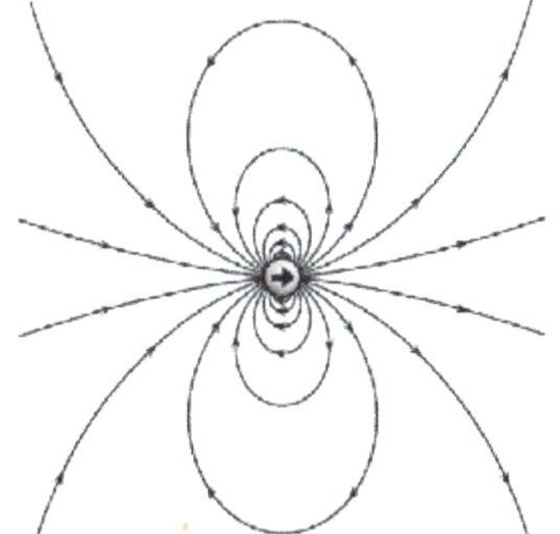

Field lines of a point dipole of any type, electric, magnetic, acoustic, etc.

A *physical dipole* consists of two equal and opposite point charges: in the literal sense, two poles.

Its field at large distances (i.e., distances large in comparison to the separation of the poles) depends almost entirely on the dipole moment as defined above. A *point (electric) dipole* is the limit obtained by letting the separation tend to 0 while keeping the dipole moment fixed. The field of a point dipole has a particularly simple form, and the order-1 term in the multipole expansion is precisely the point dipole field.

Although there are no known magnetic monopoles in nature, there are magnetic dipoles in the form of the quantum-mechanical spin associated with particles such as electrons (although the accurate description of such effects falls outside of classical electromagnetism). A theoretical magnetic *point dipole* has a magnetic field of exactly the same form as the electric field of an electric point dipole. A very small current-carrying loop is approximately a magnetic point dipole; the magnetic dipole moment of such a loop is the product of the current flowing in the loop and the (vector) area of the loop.

Any configuration of charges or currents has a 'dipole moment', which describes the dipole whose field is the best approximation, at large distances, to that of the given configuration. This is simply one term in the multipole expansion when the total charge ("monopole moment") is 0—as it *always* is for the magnetic case, since there are no magnetic monopoles. The dipole term is the dominant one at large distances: Its field falls off in proportion to $\frac{1}{r^3}$, as compared to $\frac{1}{r^4}$ for the next (quadrupole) term and higher powers of $\frac{1}{r}$ for higher terms, or $\frac{1}{r^2}$ for the monopole term.

Molecular Dipoles

Many molecules have such dipole moments due to non-uniform distributions of positive and negative charges on the various atoms. Such is the case with polar compounds like hydrogen fluoride (HF), where electron density is shared unequally between atoms. Therefore, a molecule's dipole is an electric dipole with an inherent electric field which should not be confused with a magnetic dipole which generates a magnetic field.

The physical chemist Peter J. W. Debye was the first scientist to study molecular dipoles extensively, and, as a consequence, dipole moments are measured in units named *debye* in his honor.

For molecules there are three types of dipoles:

- Permanent dipoles: These occur when two atoms in a molecule have substantially different electronegativity: One atom attracts electrons more than another, becoming more negative, while the other atom becomes more positive. A molecule with a permanent dipole moment is called a *polar* molecule.
- Instantaneous dipoles: These occur due to chance when electrons happen to be more concentrated in one place than another in a molecule, creating a temporary dipole.
- Induced dipoles: These can occur when one molecule with a permanent dipole repels another molecule's electrons, *inducing* a dipole moment in that molecule. A molecule is *polarized* when it carries an induced dipole.

More generally, an induced dipole of *any* polarizable charge distribution ρ (remember that a molecule has a charge distribution) is caused by an electric field external to ρ. This field may, for instance, originate from an ion or polar molecule in the vicinity of ρ or may be macroscopic (e.g., a molecule between the plates of a charged capacitor). The size of the induced dipole moment is equal to the product of the strength of the external field and the dipole polarizability of ρ.

Dipole moment values can be obtained from measurement of the dielectric constant. Some typical gas phase values in debye units are:

- carbon dioxide: 0
- carbon monoxide: 0.112 D
- ozone: 0.53 D
- phosgene: 1.17 D
- water vapor: 1.85 D
- hydrogen cyanide: 2.98 D
- cyanamide: 4.27 D
- potassium bromide: 10.41 D

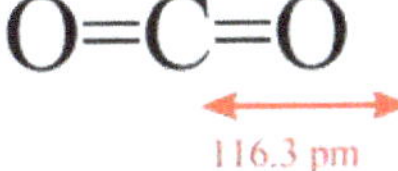

The linear molecule CO_2 has a zero dipole as the two bond dipoles cancel.

KBr has one of the highest dipole moments because it is an ionic compound that exists as a molecule in the gas phase.

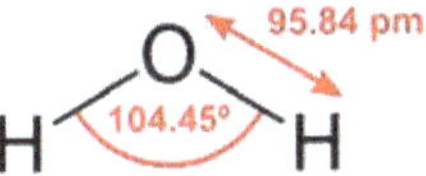

The bent molecule H_2O has a net dipole. The two bond dipoles do not cancel.

The overall dipole moment of a molecule may be approximated as a vector sum of bond dipole moments. As a vector sum it depends on the relative orientation of the bonds, so that from the dipole moment information can be deduced about the molecular geometry.

For example, the zero dipole of CO_2 implies that the two C=O bond dipole moments cancel so that the molecule must be linear. For H_2O the O–H bond moments do not cancel because the molecule is bent. For ozone (O_3) which is also a bent molecule, the bond dipole moments are not zero even though the O–O bonds are between similar atoms. This agrees with the Lewis structures for the resonance forms of ozone which show a positive charge on the central oxygen atom.

An example in organic chemistry of the role of geometry in determining dipole moment is the *cis*

and *trans* isomers of 1,2-dichloroethene. In the *cis* isomer the two polar C–Cl bonds are on the same side of the C=C double bond and the molecular dipole moment is 1.90 D. In the *trans* isomer, the dipole moment is zero because the two C–Cl bonds are on opposite sides of the C=C and cancel (and the two bond moments for the much less polar C–H bonds also cancel).

```
H     H
 \   /
  C=C
 /   \
Cl    Cl
```

Cis isomer, dipole moment 1.90 D

```
H     Cl
 \   /
  C=C
 /   \
Cl    H
```

Trans isomer, dipole moment zero

Another example of the role of molecular geometry is boron trifluoride, which has three polar bonds with a difference in electronegativity greater than the traditionally cited threshold of 1.7 for ionic bonding. However, due to the equilateral triangular distribution of the fluoride ions about the boron cation center, the molecule as a whole does not exhibit any identifiable pole: one cannot construct a plane that divides the molecule into a net negative part and a net positive part.

Quantum Mechanical Dipole Operator

Consider a collection of N particles with charges q_i and position vectors r_i. For instance, this collection may be a molecule consisting of electrons, all with charge $-e$, and nuclei with charge eZ_i, where Z_i is the atomic number of the ith nucleus. The dipole observable (physical quantity) has the quantum mechanical dipole operator:

$$\mathfrak{p} = \sum_{i=1}^{N} q_i \mathbf{r}_i .$$

Notice that this definition is valid only for non-charged dipoles, i.e. total charge equal to zero. To a charged dipole we have the next equation:

$$\mathfrak{p} = \sum_{i=1}^{N} q_i (\mathbf{r}_i - \mathbf{r}_c).$$

where r_c is the center of mass of the molecule/group of particles.

Atomic Dipoles

A non-degenerate (S-state) atom can have only a zero permanent dipole. This fact follows quantum mechanically from the inversion symmetry of atoms. All 3 components of the dipole operator are antisymmetric under inversion with respect to the nucleus,

$$\mathfrak{I}\,\mathfrak{p}\,\mathfrak{I}^{-1} = -\mathfrak{p},$$

where $\mathfrak{p}$ is the dipole operator and $\mathfrak{I}$ is the inversion operator. The permanent dipole moment of an atom in a non-degenerate state is given as the expectation (average) value of the dipole operator,

$$\langle \mathfrak{p} \rangle = \langle S|\,\mathfrak{p}\,|S\rangle,$$

where $|S\rangle$ is an S-state, non-degenerate, wavefunction, which is symmetric or antisymmetric under inversion: $\mathfrak{I}|S\rangle = \pm|S\rangle$. Since the product of the wavefunction (in the ket) and its complex conjugate (in the bra) is always symmetric under inversion and its inverse,

$$\langle \mathfrak{p} \rangle = \langle \mathfrak{I}^{-1} S|\,\mathfrak{p}\,|\mathfrak{I}^{-1} S\rangle = \langle S|\,\mathfrak{I}\mathfrak{p}\mathfrak{I}^{-1}\,|S\rangle = -\langle \mathfrak{p} \rangle$$

it follows that the expectation value changes sign under inversion. We used here the fact that $\mathfrak{I}$, being a symmetry operator, is unitary: $\mathfrak{I}^{-1} = \mathfrak{I}^{*}$ and by definition the Hermitian adjoint $\mathfrak{I}^{*}$ may be moved from bra to ket and then becomes $\mathfrak{I}^{**} = \mathfrak{I}$. Since the only quantity that is equal to minus itself is the zero, the expectation value vanishes,

$$\langle \mathfrak{p} \rangle = 0.$$

In the case of open-shell atoms with degenerate energy levels, one could define a dipole moment by the aid of the first-order Stark effect. This gives a non-vanishing dipole (by definition proportional to a non-vanishing first-order Stark shift) only if some of the wavefunctions belonging to the degenerate energies have opposite parity; i.e., have different behavior under inversion. This is a rare occurrence, but happens for the excited H-atom, where 2s and 2p states are "accidentally" degenerate and have opposite parity (2s is even and 2p is odd).

Field of a Static Magnetic Dipole

Magnitude

The far-field strength, B, of a dipole magnetic field is given by

$$B(m,r,\lambda) = \frac{\mu_0}{4\pi}\frac{m}{r^3}\sqrt{1+3\sin^2\lambda},$$

where

B is the strength of the field, measured in teslas

r is the distance from the center, measured in metres

λ is the magnetic latitude (equal to 90° – θ) where θ is the magnetic colatitude, measured in radians or degrees from the dipole axis

m is the dipole moment (VADM=virtual axial dipole moment), measured in ampere-square metres ($A{\cdot}m^2$), which equals joules per tesla

μ_0 is the permeability of free space, measured in henries per metre.

Conversion to cylindrical coordinates is achieved using $r^2 = z^2 + \rho^2$ and

$$\lambda = \arcsin\left(\frac{z}{\sqrt{z^2+\rho^2}}\right)$$

where ρ is the perpendicular distance from the z-axis. Then,

$$B(\rho,z) = \frac{\mu_0 m}{4\pi\left(z^2+\rho^2\right)^{\frac{3}{2}}}\sqrt{1+\frac{3z^2}{z^2+\rho^2}}$$

Vector Form

The field itself is a vector quantity:

$$\mathbf{B}(\mathbf{m},\mathbf{r}) = \frac{\mu_0}{4\pi}\left(\frac{3(\mathbf{m}\cdot\hat{\mathbf{r}})\hat{\mathbf{r}}-\mathbf{m}}{r^3}\right) + \frac{2\mu_0}{3}\mathbf{m}\delta^3(\mathbf{r})$$

where

B is the field

r is the vector from the position of the dipole to the position where the field is being measured

r is the absolute value of r: the distance from the dipole

$\hat{r} = \frac{r}{r}$ is the unit vector parallel to r;

m is the (vector) dipole moment

μ_o is the permeability of free space

δ^3 is the three-dimensional delta function.

This is *exactly* the field of a point dipole, *exactly* the dipole term in the multipole expansion of an arbitrary field, and *approximately* the field of any dipole-like configuration at large distances.

Magnetic Vector Potential

The vector potential A of a magnetic dipole is

$$\mathbf{A}(\mathbf{r}) = \frac{\mu_0}{4\pi}\frac{\mathbf{m}\times\hat{\mathbf{r}}}{r^2}$$

with the same definitions as above.

Field from an Electric Dipole

The electrostatic potential at position r due to an electric dipole at the origin is given by:

$$\Phi(\mathbf{r}) = \frac{1}{4\pi\varepsilon_0}\frac{\mathbf{p}\cdot\hat{\mathbf{r}}}{r^2}$$

where

$\hat{\mathbf{r}}$ is a unit vector in the direction of r, p is the (vector) dipole moment, and ε_0 is the permittivity of free space.

This term appears as the second term in the multipole expansion of an arbitrary electrostatic potential Φ(r). If the source of Φ(r) is a dipole, as it is assumed here, this term is the only non-vanishing term in the multipole expansion of Φ(r). The electric field from a dipole can be found from the gradient of this potential:

$$\mathbf{E} = -\nabla\Phi = \frac{1}{4\pi\epsilon_0}\left(\frac{3(\mathbf{p}\cdot\hat{\mathbf{r}})\hat{\mathbf{r}} - \mathbf{p}}{r^3}\right) - \frac{1}{3\epsilon_0}\mathbf{p}\delta^3(\mathbf{r})$$

where E is the electric field and δ^3 is the 3-dimensional delta function. This is formally identical to the magnetic H field of a point magnetic dipole with only a few names changed.

Torque on a Dipole

Since the direction of an electric field is defined as the direction of the force on a positive charge, electric field lines point away from a positive charge and toward a negative charge.

When placed in an electric or magnetic field, equal but opposite forces arise on each side of the dipole creating a torque τ:

$$\tau = \mathbf{p}\times\mathbf{E}$$

for an electric dipole moment p (in coulomb-meters), or

$$\tau = \mathbf{m}\times\mathbf{B}$$

for a magnetic dipole moment m (in ampere-square meters).

The resulting torque will tend to align the dipole with the applied field, which in the case of an electric dipole, yields a potential energy of

$$U = -\mathbf{p}\cdot\mathbf{E}.$$

The energy of a magnetic dipole is similarly

$$U = -\mathbf{m}\cdot\mathbf{B}.$$

Dipole Radiation

In addition to dipoles in electrostatics, it is also common to consider an electric or magnetic dipole that is oscillating in time. It is an extension, or a more physical next-step, to spherical wave radiation.

In particular, consider a harmonically oscillating electric dipole, with angular frequency ω and a dipole moment p_0 along the □ direction of the form

$$\mathbf{p}(\mathbf{r},t) = \mathbf{p}(\mathbf{r})e^{-i\omega t} = p_0\hat{\mathbf{z}}e^{-i\omega t}.$$

In vacuum, the exact field produced by this oscillating dipole can be derived using the retarded potential formulation as:

$$\mathbf{E} = \frac{1}{4\pi\varepsilon_0}\left\{\frac{\omega^2}{c^2 r}(\hat{\mathbf{r}}\times\mathbf{p})\times\hat{\mathbf{r}} + \left(\frac{1}{r^3} - \frac{i\omega}{cr^2}\right)\left[3\hat{\mathbf{r}}(\hat{\mathbf{r}}\cdot\mathbf{p}) - \mathbf{p}\right]\right\}e^{i\omega r/c}e^{-i\omega t}$$

$$\mathbf{B} = \frac{\omega^2}{4\pi\varepsilon_0 c^3}\hat{\mathbf{r}}\times\mathbf{p}\left(1 - \frac{c}{i\omega r}\right)\frac{e^{i\omega r/c}}{r}e^{-i\omega t}.$$

For $\frac{r\omega}{c} \gg 1$, the far-field takes the simpler form of a radiating "spherical" wave, but with angular dependence embedded in the cross-product:

$$\mathbf{B} = \frac{\omega^2}{4\pi\varepsilon_0 c^3}(\hat{\mathbf{r}}\times\mathbf{p})\frac{e^{i\omega(r/c-t)}}{r} = \frac{\omega^2\mu_0 p_0}{4\pi c}(\hat{\mathbf{r}}\times\hat{\mathbf{z}})\frac{e^{i\omega(r/c-t)}}{r} = -\frac{\omega^2\mu_0 p_0}{4\pi c}\sin\theta\frac{e^{i\omega(r/c-t)}}{r}\hat{\phi}$$

$$\mathbf{E} = c\mathbf{B}\times\hat{\mathbf{r}} = -\frac{\omega^2\mu_0 p_0}{4\pi}\sin\theta(\hat{\phi}\times\hat{\mathbf{r}})\frac{e^{i\omega(r/c-t)}}{r} = -\frac{\omega^2\mu_0 p_0}{4\pi}\sin\theta\frac{e^{i\omega(r/c-t)}}{r}\hat{\theta}.$$

The time-averaged Poynting vector

$$\langle\mathbf{S}\rangle = \left(\frac{\mu_0 p_0^2\omega^4}{32\pi^2 c}\right)\frac{\sin^2\theta}{r^2}\hat{\mathbf{r}}$$

is not distributed isotropically, but concentrated around the directions lying perpendicular to the dipole moment, as a result of the non-spherical electric and magnetic waves. In fact, the spherical harmonic function ($\sin\theta$) responsible for such toroidal angular distribution is precisely the $l = 1$ "p" wave.

The total time-average power radiated by the field can then be derived from the Poynting vector as

$$P = \frac{\mu_0\omega^4 p_0^2}{12\pi c}.$$

Notice that the dependence of the power on the fourth power of the frequency of the radiation is in accordance with the Rayleigh scattering, and the underlying effects why the sky consists of mainly blue colour.

A circular polarized dipole is described as a superposition of two linear dipoles.

Linear Polarization

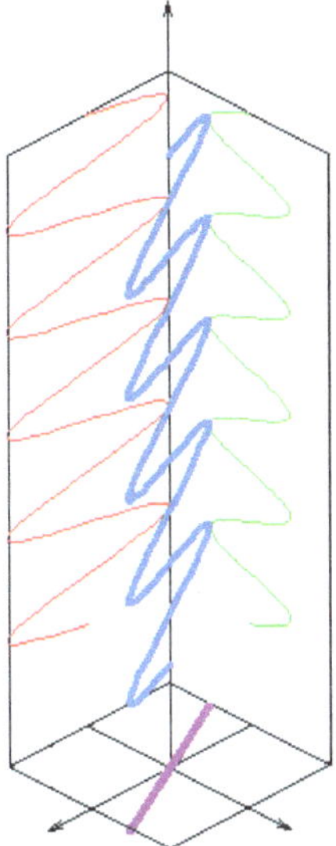

Diagram of the electric field of a light wave (blue), linear-polarized along a plane (purple line), and consisting of two orthogonal, in-phase components (red and green waves)

In electrodynamics, linear polarization or plane polarization of electromagnetic radiation is a confinement of the electric field vector or magnetic field vector to a given plane along the direction of propagation.

The orientation of a linearly polarized electromagnetic wave is defined by the direction of the electric field vector. For example, if the electric field vector is vertical (alternately up and down as the wave travels) the radiation is said to be vertically polarized.

Mathematical Description of Linear Polarization

The classical sinusoidal plane wave solution of the electromagnetic wave equation for the electric and magnetic fields is (cgs units)

$$\mathbf{E}(\mathbf{r},t) = |\mathbf{E}| \operatorname{Re}\left\{|\psi\rangle \exp\left[i\left(kz-\omega t\right)\right]\right\}$$

$$\mathbf{B}(\mathbf{r},t) = \hat{\mathbf{z}} \times \mathbf{E}(\mathbf{r},t)/c$$

for the magnetic field, where k is the wavenumber,

$$\omega = ck$$

is the angular frequency of the wave, and c is the speed of light.

Here $|\mathbf{E}|$ is the amplitude of the field and

$$|\psi\rangle \overset{\text{def}}{=} \begin{pmatrix} \psi_x \\ \psi_y \end{pmatrix} = \begin{pmatrix} \cos\theta \exp(i\alpha_x) \\ \sin\theta \exp(i\alpha_y) \end{pmatrix}$$

is the Jones vector in the x-y plane.

The wave is linearly polarized when the phase angles α_x, α_y are equal,

$$\alpha_x = \alpha_y \overset{\text{def}}{=} \alpha.$$

This represents a wave polarized at an angle θ with respect to the x axis. In that case, the Jones vector can be written

$$|\psi\rangle = \begin{pmatrix} \cos\theta \\ \sin\theta \end{pmatrix} \exp(i\alpha).$$

The state vectors for linear polarization in x or y are special cases of this state vector.

If unit vectors are defined such that

$$|x\rangle \overset{\text{def}}{=} \begin{pmatrix} 1 \\ 0 \end{pmatrix}$$

and

$$|y\rangle \overset{\text{def}}{=} \begin{pmatrix} 0 \\ 1 \end{pmatrix}$$

then the polarization state can be written in the "x-y basis" as

$$|\psi\rangle = \cos\theta \exp(i\alpha)|x\rangle + \sin\theta \exp(i\alpha)|y\rangle = \psi_x|x\rangle + \psi_y|y\rangle.$$

Vector Nature of Electromagnetic Radiation

Consider a situation where the same electrical signal is fed to two mutually perpendicular dipoles, one along the y axis and another along the z axis as shown in Figure.

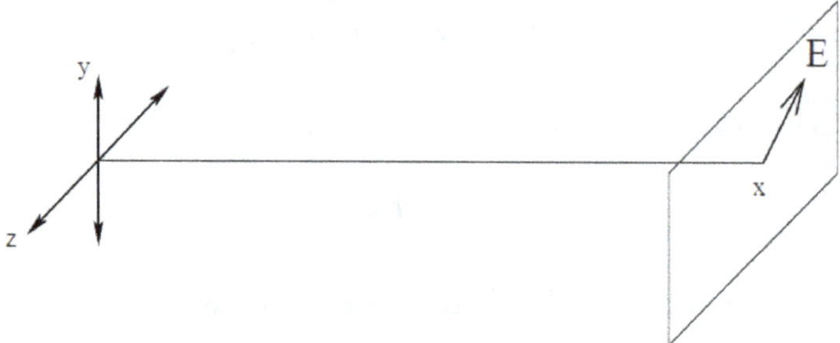

The resultant electric field at x

We are interested in the electric field at a distant point along the x axis. The electric field is a superposition of two components

$$\vec{E}(x,t) = E_y(x,t)\hat{j} + E_z(x,t)\hat{k}$$

one along the y axis produced by the dipole which is aligned along the y axis, and another along the z axis produced by the dipole oriented along the z axis.

Linear Polarization

In this situation where both dipoles receive the same signal, the two components are equal $E_y = E_z$ and

$$\vec{E}(x,t) = E(\hat{j} + \hat{k})\cos(\omega t - kx)$$

If we plot the time evolution of the electric field at a fixed position we see that it oscillates up and down along a direction which is at 45° to the y and z axis.

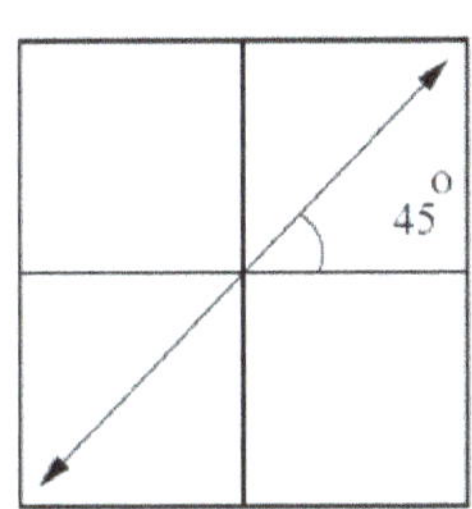

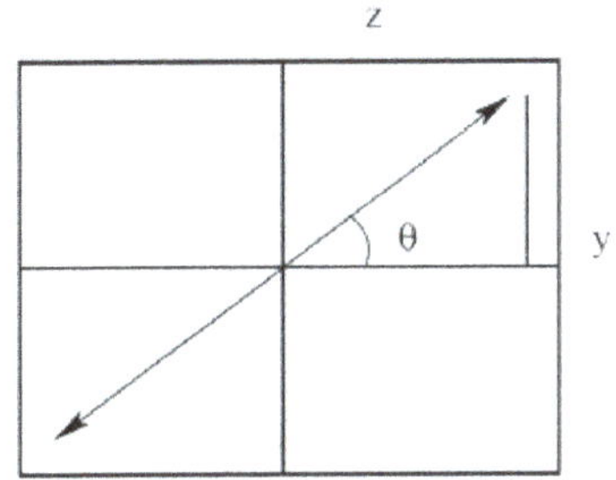

Linear Polarization

The point to note is that it is possible to change the relative amplitudes of E_y *and* E_z by changing the currents in the oscillators. The resultant electric field is

$$\vec{E}(x,t) = (E_y\hat{k} + E_z\hat{j})\cos(\omega t - kx)$$

The resultant electric field vector has magnitude $E = \sqrt{E_y^2 + E_z^2}$ and it oscillates along a direction at an angle $\theta = \tan^{-1}\left(\frac{E_y}{E_z}\right)$ with respect to the z axis.

Under no circumstance does the electric field have a component along the direction of the wave i.e along the x axis. The electric field can be oriented along any direction in the $y-z$ plane. In the cases which we have considered until now, the electric field oscillates up and down a fixed direction in the $y-z$ plane. Such an electromagnetic wave is said to be linearly polarized.

Circular Polarization

The polarization of the wave refers to the time evolution of the electric field vector. An interesting situation occurs if the same signal is fed to the two dipoles, but the signal to the z axis is given an extra $\pi/2$ phase. The electric field now is

$$\vec{E}(x,t) = E\left[\cos(\omega t - kx)\hat{j} + \cos(\omega t - kx + \pi/2)\hat{k}\right]$$

$$= E\left[\cos(\omega t - kx)\hat{j} - \sin(\omega t - kx)\hat{k}\right]$$

If we now follow the evolution of $\vec{E}(t)$ at a fixed point, we see that the tip of the vector $\vec{E}(t)$ moves on a circle of radius E clockwise (when the observer looks towards the source) as shown in Figure.

We call such a wave right circularly polarized. The electric field would have rotated in the opposite direction had we applied a phase lag of $\pi/2$. We would then have obtained a left circularly polarized wave.

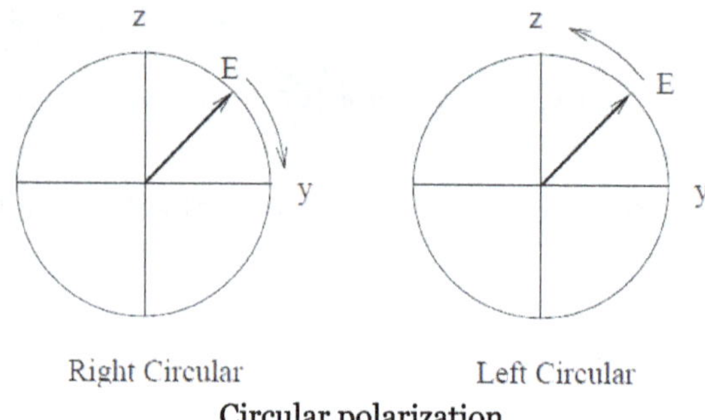

Circular polarization

Elliptical Polarization.

Oscillations of different amplitude combined with a phase difference of $\pi/2$ produces elliptically polarized wave where the ellipse is aligned with the $y-z$ axis as shown in Figure. The ellipse is not aligned with the $y-z$ axis for an arbitrary phase difference between the y and z components of the electric field. This is the most general state of polarization shown in the last diagram of the Figure. Linear and circularly polarized waves are specific cases of Elliptically polarized waves.

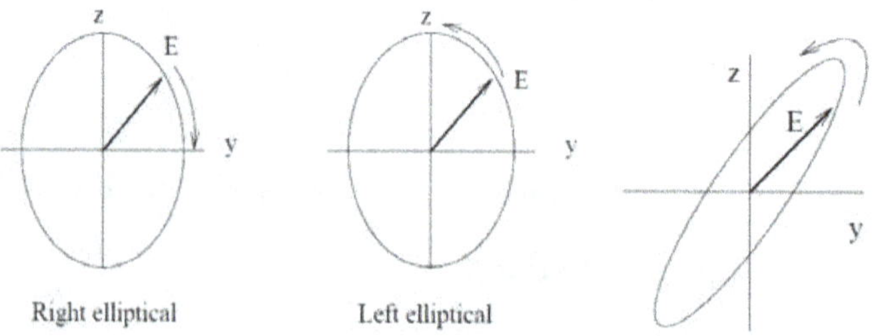

Elliptical polarization

Spectrum of Electromagnetic Radiation

Electromagnetic waves come in a wide range of frequency or equivalently wavelength. We refer to different bands of the frequency spectrum using different names. These bands are often overlapping. The nomenclature of the bands and the frequency range are typically based on the properties of the waves and the techniques to generate or detect them.

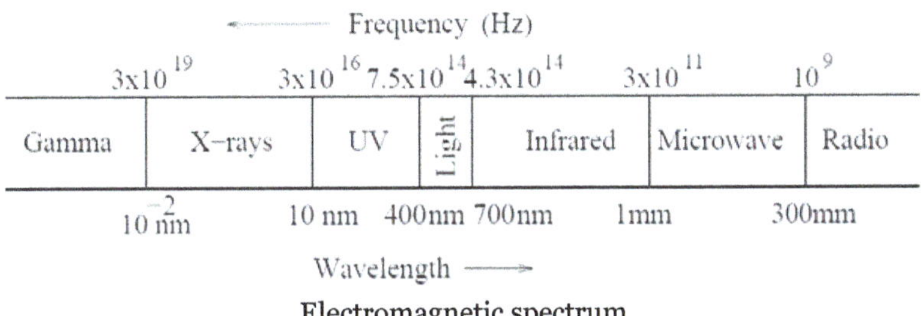

Electromagnetic spectrum

Radiowave and Microwave

Radiowaves span frequencies less than 1GHz. The higher frequency part of this band is used for radio and television transmission. There is no theoretical lower limit, and the familiar 50Hz power supply lines emits radio waves at this frequency which corresponds to a wavelength.

$$\lambda = \frac{c}{v} = \frac{3\times10^{8}ms^{-1}}{50s^{-1}} = 6\times10^{6}m$$

The upper frequency part of this band, $VHF(30\ MHz - 300\ MHz)$ and $UHF(300\ MHz - 3\ GHz)$, is used for FM and television transmission whereas

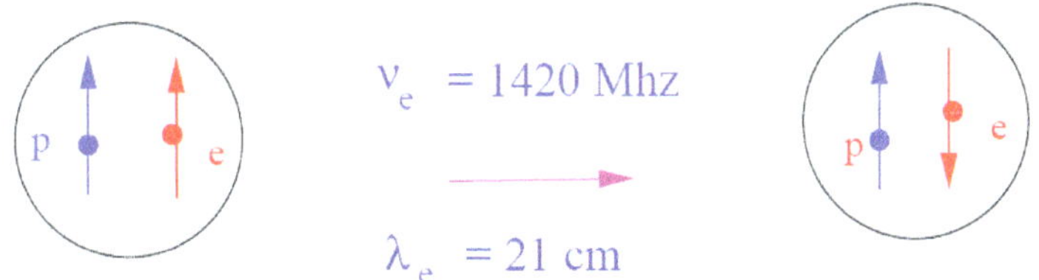

Hyperfine splitting in neurtral hydrogen

the medium waves $(0.5\ MHz - 1.5\ MHz)$ and short waves $(3\ MHz - 30\ MHz)$ are used for radio transmission.

Microwave refers to electromagnetic waves in the frequency range 1GHz to 300GHz or wavelengths 30cm to 1mm. The communication band extends into the microwave.

The communication band extends into the microwave. Global system for mobile (GSM) operates in 900/1800/1900/MHz bands.

The earths atmosphere is largely transparent to electromagnetic waves from 1cm to 30cm. Consequently, radiowaves and microwaves are both very useful for space communication and astronomy. This branch of astronomy is called radio astronomy.

In these two bands there are a large variety of electrical circuits and antennas that are used to produce waves for communication. We briefly discuss below a few of the astronomical sources of radio and microwave radiation.

21cm Radiation

Hydrogen is the most abundant element in the universe. Atomic neutral hydrogen, has two states, one where the proton and electron spins are aligned and another where they are opposite as shown in Figure. The separation between these two energy states is known as hyperfine splitting in hydrogen. A transition between these two states causes radiation at 1.42GHz or 21cm to be emitted. This is a very important source of information about our Galaxy and external galaxies, neutral hydrogen being found in many galaxies including our own. The Giant Meterwave Radio Telescope located in Narayangaon near Pune is currently the worlds largest low frequency radio telescope operating in several frequency bands from 1.42GHz to 50MHz. Figure also shows the image of a dwarf galaxy DDO210 made with the GMRT using the 21cm radiation from neutral hydrogen. The contours show how the neutral hydrogen is distributed while the distribution of stars is shown in black. It is clear that the hydrogen gas which is referred to as the interstellar medium is spread out over a much larger region compared to the stars.

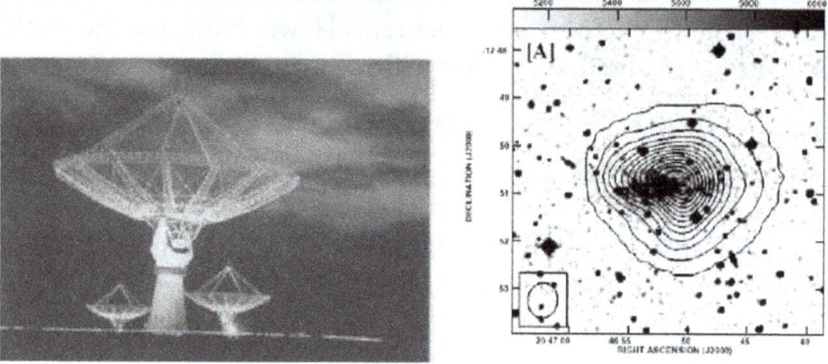

GMRT and radio image of a dwarf galaxy

Cosmic Microwave Background Radiation

An object which is equally efficient in absorbing and emitting radiation of all frequencies is referred to as a black body. Consider a cavity enclosed inside a black body at a temperature T. The electromagnetic waves inside this cavity will be repeatedly absorbed and re-emitted by the walls of this cavity until the radiation is in thermal equilibrium with the black body. It is found that the radiation spectum inside this black body cavity is completely specified by T the temperature of the black body. This radiation is referred to as black body radiation. Writing the energy density du_ν of the black body radiation in a frequency interval $d\nu$ as

$$du_\nu = u_\nu \, d\nu,$$

the spectral energy density u_ν is found to be given by

$$u_\nu = \frac{8\pi h\nu^3}{c^3} \frac{1}{\left[\exp\left(\frac{h\nu}{kT}\right) - 1\right]},$$

where $h =$ 6.63×10^{-34} Joule-sec is the Planck constant and $k = R/N = 8.314/(6.022\times10^{23})$ = 1.38×10^{-23} Joules/Kelvin is the Boltzmann Constant. The spectral energy density can equivalently be defines in terms of the wavelength interval $d\lambda$ as $du\lambda$ = $u_\lambda \, d\lambda$. Figure (a) shows the energy density of black body radiation for different values of the temperature T. The curves for different temperatures are unique and the curves corresponding to different values of T do not intersect.

The wavelength λ_m at which the energy density peaks decreases with T, and the relation is given by the Wien's displacement law

$$\lambda_m T = 2.898 \times 10^{-3} m\ K$$

Radio observations carried out by pointing a radio receiver in different directions on the sky show that there is a radiation with a black body spectrum

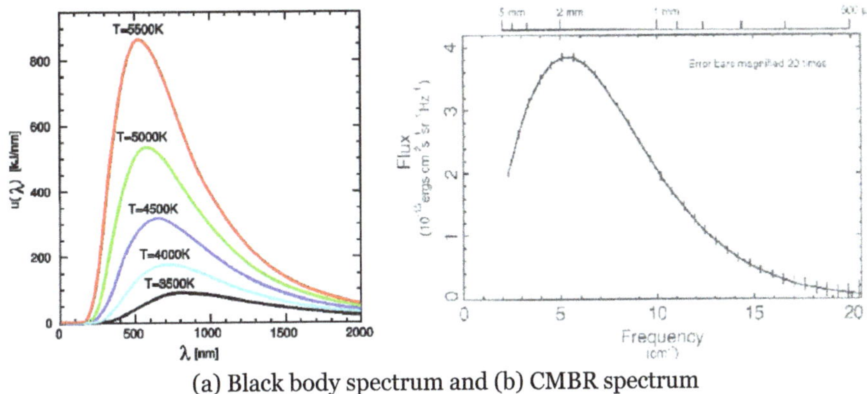

(a) Black body spectrum and (b) CMBR spectrum

(Figure (b)) at $T = 2.735 \pm 0.06K$ arriving from all directions in the sky. This radiation is not terrestrial in origin. It is believed that we are actually seeing a radiation which pervades the whole universe and is a relic of a hot past referred to as the hot Big Bang. This black body radiation is referred to as the Cosmic Microwave Background Radiation (CMBR) which peaks in the microwave region of the Spectrum.

Molecular Lines

H^+ p H^+

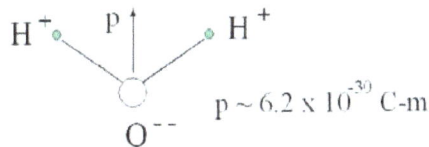

$p \sim 6.2 \times 10^{-30}$ C-m

O^{--}

Polar molecules like water are influenced by any incident electromagnetic radiation, and such molecules can be set into vibrations or rotation by the external, oscillating electric field. The water molecules try to align their dipole movement with the external electric field which is oscillating. This sets themolecule into rotation. The rotational energy levels are quantized, and the molecules have distinct frequencies at which there are resonances where the molecule emits or absorbs maximum energy. The Microwave ovens utilize a 2.45GHz rotational transition of water. The water molecules absorb the incident electromagnetic radiation and start rotating. The rotational energy of the molecules is converted to random motions or heat.

Most of the rotational and vibrational transitions of molecules lie in the microwave and Infrared (IR) bands. The frequency range 50GHz to 10THz is often called Tera hertz radiation or T-rays. The water vapour in the atmosphere is opaque at much of these frequencies. Dry substances like paper, plastic which do not have water molecules are transparent to T- rays, whereas it is absorbed by substances with water and is reflected by metals which makes it suitable for various imaging applications.

Infrared

The frequency range from $3\times10^{11}\,Hz\ to \sim 4\times10^{14}\ Hz$ is referred to as the infrared band (IR). The frequency is just below that of red light. The IR band is again subdivided as follows

near IR	780 - 3000 nm,
intermediate IR	3000 - 6000 nm,
far IR	6000 - 15000 nm,
extreme IR	15000 nm - 1mm.

The divisions are quite loose and they vary.

There are many molecular vibrational and rotational transitions which produce radiation in the IR. Vibrational transitions of $CO_2\ \ and\ \ H_2O$ fall in the energy $0.2-0.8\ eV$ range. The energy of a photon is calculated by multiplying its frequency with Planck constant h and converting the Joules in electron volts $(eV)(1eV = 1.6\times10^{-19}\,Joules)$. Usually small wavelength (or the large frequency) bands of the electromagnetic radiation are distinguished by their energies. Further, the black body radiation from hot objects emit copiousamounts of IR as the spectrum typically peaks in this band.

For example the black body radiation from human beings peaks at around 10, 000 nm asking the IR suitable for "night vision". Some snakes sense their preys using infrared vision in night. Stars like the Sun and incandescent lamps emit copiously in the IR. In fact an incandescent lamp radiates away more than 50% of its energy in the IR.

IR spy satellites are used to monitor events of sudden heat generation indicating a rocket launch or possibly nuclear explosion.

Much of fiber optical communication also works in the IR.

Visible Light

This corresponds to electromagnetic radiation in the wavelength range $3.84\times10^{14}\,Hz\ to\ 7.69\times10^{14}\,Hz.$ .The atmosphere is transparent in this band, which is possibly why we have developed vision in this frequency range.

Visible radiation is produced when electrons in the outer shell of atoms jump from a higher energy level to a lower lenergy level. These are used in sources of visible radiation as well as detectors. Black body radiation from objects at a temperature of a few thousand Kelvin is another source of visible light.

Colour is our perception of the relative contribution from different frequencies. Different combinations of the frequencies can produce the same colour.

Ultraviolet (UV)

These are electromagnetic waves with frequency just higher than blue light, and this band spans from $8\times10^{14}\ Hz\ to\ 3.4\times10^{16}\,Hz$. Typical energy range corresponds to 3–100 eV.

One of the main sources of UV radiation is the Sun. UV radiation has sufficient energy to ionize the atoms in the upper atmosphere, producing the ionosphere. Exposure to UV radiation is harmful to living organisms, and it is used to kill bacteria in many water purification systems. Wavelenghs less than 300 nm are well suited for this killing. Fortunately Ozone (O_3) in the atmosphere absorbs UV radiation protecting life on the earth's surface.

The human eyes cannot see in UV as it is absorbed by the lens. Peoplewhose lens have been removed due to cataract can see in UV to some extant. Many insects can see in part of the UV band.

UV is produced through more energetic electronic transitions in atoms. For example the transition of the hydrogen atom from the first excited state to the ground slate (Lyman-α) produces UV.

UV is used in lithography for making ICS, erasing EPROMS, etc.

X-rays

Electromagnetic waves with frequency in the range $2.4\times10^{16}\,Hz\ to\ 5\times10^{19}\,Hz$ is referred to as X-ray. The energy range of the X-rays are between 0.1 and 200 KeV. These are produced by very fast moving electrons when they encounter positively charged nuclei of atoms and are accelerated as a consequence. This occurs when electrons are bombarded on a copper plate which is typically how X-rays are produced. Figure shows a X-ray tube where electrons are accelerated by a voltage in the range 30 to 150KeV and then bombarded onto a copper plate.

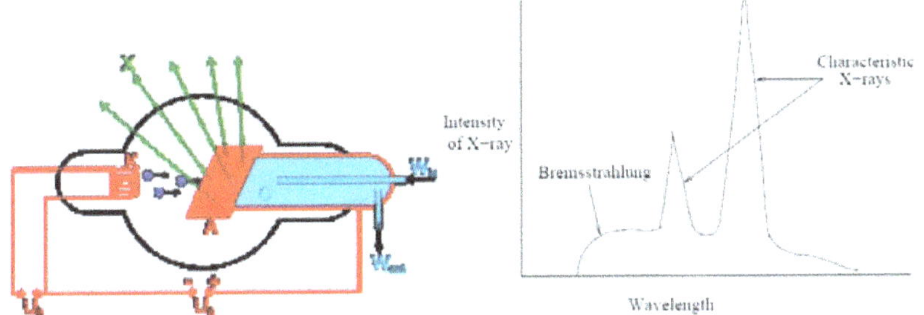

(a) An X-ray tube (b) Characteristic X-rays and Bremsstrahlung

Characteristic X-rays are produced through inner shell electron transitions in atoms. Characteristic X-rays are discrete in wavelengths. Different elements

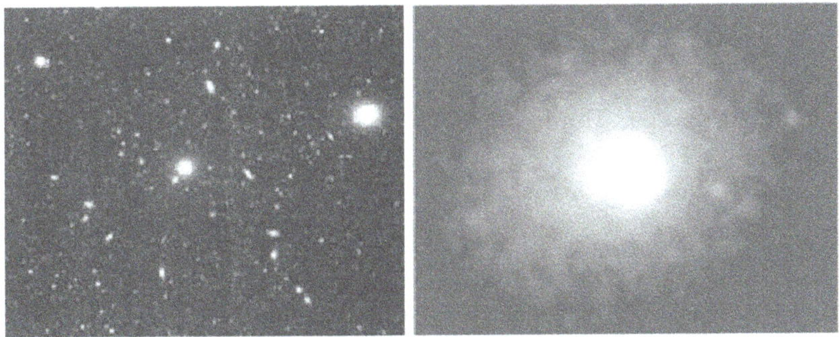

Centaurus cluster (a) Optical image (b) X-ray image

present in a substance can be identified using X-ray diffraction which will show the X-ray peaks at particular wavelengths. Accelerated or decelerated electrons produce continuous X-rays

(Bremsstrahlung). When a metal target is fired with electrons both Characteristic X-rays of the metal are produced with continuous Bremsstrahlung background.

Since the wavelengths of the X-rays are of atomic dimensions or less they are good probes for finding the structures of substances. Human body is transparent for X-rays but the bones are not. 10-100 KeV X-rays are used for diagnostic purpose to locate fractures etc.

Hot ionized gas (plasma) found in many astrophysical situations like around our sun or around black holes also produces X-ray. Figure shows the Centaurus cluster, a cluster of galaxies more than hundred in number as seen in the optical image on the left. The X-ray image of the same cluster shown on the right reveals that in addition to the galaxies there is a hot ionized gas at a temperature of few tens of million Kelvin which emits copious amounts of X-ray.

Inner shell electronic transitions in atoms also produce X-rays. X-rays are also produced in particle accelerators like the synchrotron.

Gamma Rays

The highest frequency $(> 5^{19}\,Hz)$ electromagnetic radiation is referred to as Gamma Rays. These are produced in nuclear transitions when the nucleus goes from an excited state to a lower energy state. These are produced in copious amounts in nuclear reactors and in nuclear explosions. Electron positron annihilation also produces Gamma rays. Typical energy range for Gamma radiation is in MeV (million electron-volts). They penetrate through almost any material. One needs thick lead walls to stop Gamma rays. Gamma rays will ionize most gases, and is not very difficult to detect this radiation through this. Exposure to gamma rays can mutate and even kill living cells, and is very harmful for humans. Controlled exposure is used to kill cancerous cells and this is used for cancer treatment.

There are mysterious astrophysical sources which emit a very intense burst of Gamma rays. The exact nature of these sources is still a subject of intense study, and one of the theories is that these are hypernovae in distant galaxies.

References

- Narayanan, DL; Saladi, RN; Fox, JL (September 2010). "Ultraviolet radiation and skin cancer". International Journal of Dermatology. 49 (9): 978–86. doi:10.1111/j.1365-4632.2010.04474.x. PMID 20883261
- Brau, Charles A. (2004). Modern Problems in Classical Electrodynamics. Oxford University Press. ISBN 0-19-514665-4
- Crowther, James Arnold (1920). The life and discoveries of Michael Faraday. Society for promoting Christian knowledge. pp. 54–57. Retrieved 15 June 2014
- Maxwell, J. Clerk (1 January 1865). "A Dynamical Theory of the Electromagnetic Field". Philosophical Transactions of the Royal Society of London. 155: 459–512. doi:10.1098/rstl.1865.0008
- Thorn, J. J.; Neel, M. S.; Donato, V. W.; Bergreen, G. S.; Davies, R. E.; Beck, M. (2004). "Observing the quantum behavior of light in an undergraduate laboratory" (PDF). American Journal of Physics. 72 (9): 1210. Bibcode:2004AmJPh..72.1210T. doi:10.1119/1.1737397
- Binhi, Vladimir N (2002). Magnetobiology: Underlying Physical Problems. Repiev, A & Edelev, M (translators from Russian). San Diego: Academic Press. pp. 1–16. ISBN 978-0-12-100071-4. OCLC 49700531

- Carmichael, H. J. "Einstein and the Photoelectric Effect" (PDF). Quantum Optics Theory Group, University of Auckland. Archived from the original (PDF) on 17 October 2012. Retrieved 22 December 2009
- Kinsler, P. (2010). "Optical pulse propagation with minimal approximations". Phys. Rev. A. 81: 013819. Bibcode:2010PhRvA..81a3819K. arXiv:0810.5689. doi:10.1103/PhysRevA.81.013819
- Liebel, F.; Kaur, S.; Ruvolo, E.; Kollias, N.; Southall, M. D. (2012). "Irradiation of Skin with Visible Light Induces Reactive Oxygen Species and Matrix-Degrading Enzymes". Journal of Investigative Dermatology. 132 (7): 1901–1907. PMID 22318388. doi:10.1038/jid.2011.476

4

Interference: A Phenomenon of Wave

Waves can have positive and negative values. When two waves are superposed, they can generate a wave of higher, lower or same amplitude. This phenomenon is known as interference. The aspects elucidated in this section are of vital importance, and provide a better understanding of interference.

Interference (Wave Propagation)

In physics, interference is a phenomenon in which two waves superpose to form a resultant wave of greater, lower, or the same amplitude. Interference usually refers to the interaction of waves that are correlated or coherent with each other, either because they come from the same source or because they have the same or nearly the same frequency. Interference effects can be observed with all types of waves, for example, light, radio, acoustic, surface water waves or matter waves.

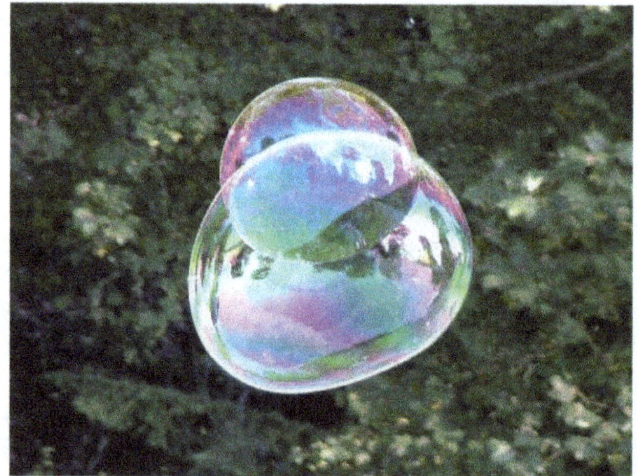

The iridescence of soap bubbles is due to thin-film interference.

Mechanisms

The principle of superposition of waves states that when two or more propagating waves of same type are incident on the same point, the resultant amplitude at that point is equal to the vector sum of the amplitudes of the individual waves. If a crest of a wave meets a crest of another wave of the same frequency at the same point, then the amplitude is the sum of the individual amplitudes—this is constructive interference. If a crest of one wave meets a trough of another wave, then the amplitude is equal to the difference in the individual amplitudes—this is known as destructive interference.

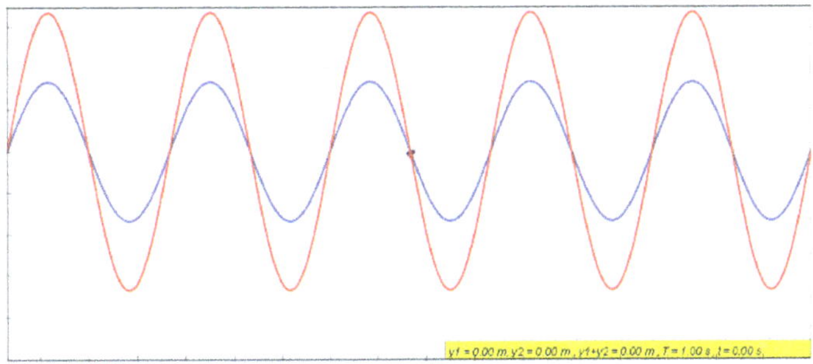

Interference of left traveling (green) and right traveling (blue) waves in one dimension, resulting in final (red) wave

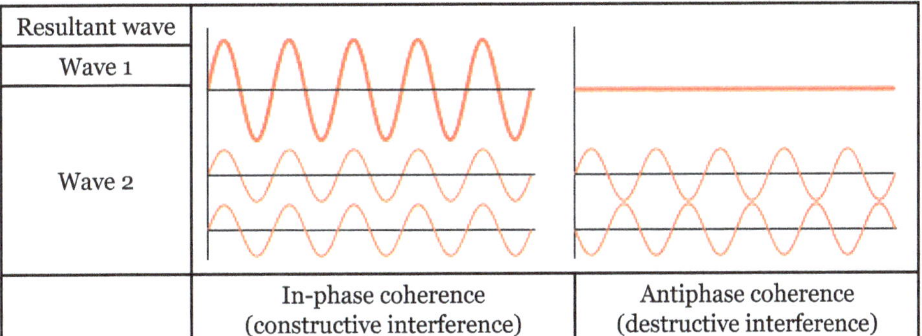

Constructive interference occurs when the phase difference between the waves is an even multiple of π (180°) (a multiple of 2π, 360°), whereas destructive interference occurs when the difference is an odd multiple of π. If the difference between the phases is intermediate between these two extremes, then the magnitude of the displacement of the summed waves lies between the minimum and maximum values.

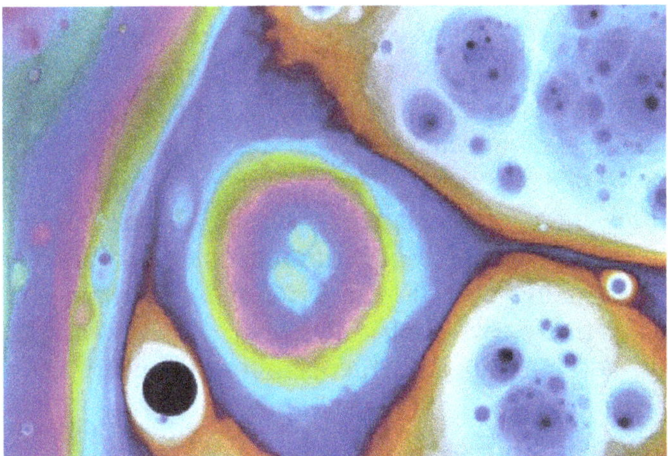

A magnified image of a coloured interference pattern in a soap film. The "black holes" are areas of almost total destructive interference, (antiphase).

Consider, for example, what happens when two identical stones are dropped into a still pool of water at different locations. Each stone generates a circular wave propagating outwards from the point where the stone was dropped. When the two waves overlap, the net displacement at a particular point is the sum of the displacements of the individual waves. At some points, these will be in phase, and will produce a maximum displacement. In other places, the waves will be in anti-phase, and there will be no net displacement at these points. Thus, parts of the surface will be stationary—these are seen in the figure above and to the right as stationary blue-green lines radiating from the centre.

Interference of waves from two point sources.

Interference of light is a common phenomenon that can be explained classically by the superposition of waves theory however a deeper understanding of light interference requires knowledge of the quantum wave propagation of light which resolves further experimental observations, see QED: The Strange Theory of Light and Matter. Prime examples of light interference are the famous Double-slit experiment, laser speckle, optical thin layers and films and interferometers. An example is the double slit experiment, classically light interferes and the energy of photons is lost, however with wave propagation the observed bright and dark areas are a result of the paths available for the photons to travel. Dark areas in the double slit are not available to the photons and bright areas are allowed paths. Laser speckle and a Michelson interferometer are examples where an observer truly observes light of differing phases, the electrons in the photo sensitive areas of the eye do not perceive photons that are out of phase with each other due to a net zero superposition of the E vector of the photons, these photons merely continue deeper into the eye tissues where they are absorbed. Thin films also behave in a quantum manner. Traditionally the classical model is taught as a basis for understanding optical interference based the Huygens–Fresnel principle and it was not until discussions in the 1920s Solvay Conference that de Broglie first proposed unique wave properties of light, Feynman made further significant contributions in the 1940s/50s and experiments continue today.

Derivation

The above can be demonstrated in one dimension by deriving the formula for the sum of two waves. The equation for the amplitude of a sinusoidal wave traveling to the right along the x-axis is

$$W_1(x,t) = A\cos(kx - \omega t)$$

where A is the peak amplitude, $k = 2\pi / \lambda$ is the wavenumber and $\omega = 2\pi f$ is the angular frequency of the wave. Suppose a second wave of the same frequency and amplitude but with a different phase is also traveling to the right

$$W_2(x,t) = A\cos(kx - \omega t + \phi)$$

where ϕ is the phase difference between the waves in radians. The two waves will superpose and add: the sum of the two waves is

$$W_1 + W_2 = A[\cos(kx - \omega t) + \cos(kx - \omega t + \phi)]$$

Using the trigonometric identity for the sum of two cosines: $\cos a + \cos b = 2\cos(\frac{a+b}{2})\cos(\frac{a-b}{2})$, this can be written

$$W_1 + W_2 = 2A\cos\left(kx - \omega t + \frac{\phi}{2}\right)\cos\left(\frac{\phi}{2}\right)$$

This represents a wave at the original frequency, traveling to the right like the components, whose amplitude is proportional to the cosine of $\phi/2$.

- *Constructive interference*: If the phase difference is an even multiple of pi: $\phi = \ldots -4\pi, -2\pi, 0, 2\pi, 4\pi, \ldots$ then $|\cos\phi/2| = 1$, so the sum of the two waves is a wave with twice the amplitude

$$W_1 + W_2 = 2A\cos(kx - \omega t)$$

- *Destructive interference*: If the phase difference is an odd multiple of pi: $\phi = \ldots -3\pi, -\pi, \pi, 3\pi, 5\pi, \ldots$ then $\cos\phi/2 = 0$, so the sum of the two waves is zero

$$W_1 + W_2 = 0$$

Between Two Plane Waves

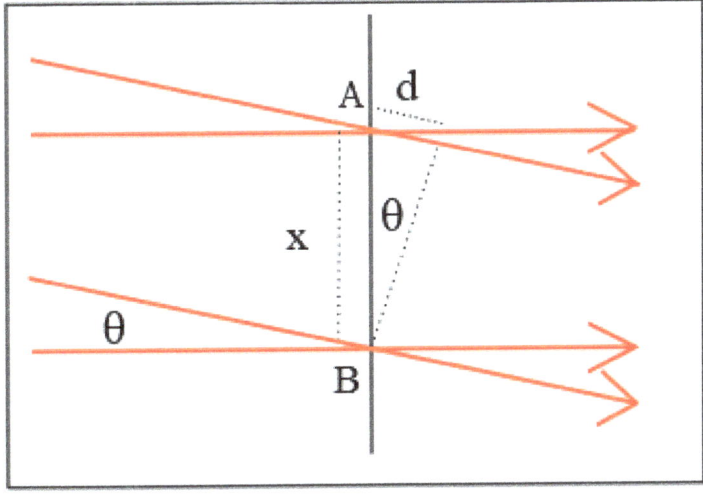

Geometrical arrangement for two plane wave interference

Interference fringes in overlapping plane waves

A simple form of interference pattern is obtained if two plane waves of the same frequency intersect at an angle. Interference is essentially an energy redistribution process. The energy which is lost at the destructive interference is regained at the constructive interference. One wave is travelling horizontally, and the other is travelling downwards at an angle θ to the first wave. Assuming that the two waves are in phase at the point B, then the relative phase changes along the x-axis. The phase difference at the point A is given by

$$\Delta\varphi = \frac{2\pi d}{\lambda} = \frac{2\pi x \sin\theta}{\lambda}$$

It can be seen that the two waves are in phase when

$$\frac{x\sin\theta}{\lambda} = 0, \pm 1, \pm 2, \ldots,$$

and are half a cycle out of phase when

$$\frac{x\sin\theta}{\lambda} = \pm\frac{1}{2}, \pm\frac{3}{2}, \ldots$$

Constructive interference occurs when the waves are in phase, and destructive interference when they are half a cycle out of phase. Thus, an interference fringe pattern is produced, where the separation of the maxima is

$$d_f = \frac{\lambda}{\sin\theta}$$

and d_f is known as the fringe spacing. The fringe spacing increases with increase in wavelength, and with decreasing angle θ.

The fringes are observed wherever the two waves overlap and the fringe spacing is uniform throughout.

Between two Spherical Waves

A point source produces a spherical wave. If the light from two point sources overlaps, the interference pattern maps out the way in which the phase difference between the two waves varies in space. This depends on the wavelength and on the separation of the point sources. The figure to the right shows interference between two spherical waves. The wavelength increases from top to bottom, and the distance between the sources increases from left to right.

When the plane of observation is far enough away, the fringe pattern will be a series of almost straight lines, since the waves will then be almost planar.

Optical interference between two point sources for different wavelengths and source separations

Multiple Beams

Interference occurs when several waves are added together provided that the phase differences between them remain constant over the observation time.

It is sometimes desirable for several waves of the same frequency and amplitude to sum to zero (that is, interfere destructively, cancel). This is the principle behind, for example, 3-phase power and the diffraction grating. In both of these cases, the result is achieved by uniform spacing of the phases.

It is easy to see that a set of waves will cancel if they have the same amplitude and their phases are spaced equally in angle. Using phasors, each wave can be represented as $Ae^{i\varphi_n}$ for N waves from $n=0$ to $n=N-1$, , where

$$\varphi_n - \varphi_{n-1} = \frac{2\pi}{N}.$$

To show that

$$\sum_{n=0}^{N-1} Ae^{i\varphi_n} = 0$$

one merely assumes the converse, then multiplies both sides by $e^{i\frac{2\pi}{N}}$

The Fabry–Pérot interferometer uses interference between multiple reflections.

A diffraction grating can be considered to be a multiple-beam interferometer, since the peaks which it produces are generated by interference between the light transmitted by each of the elements in the grating.

Optical Interference

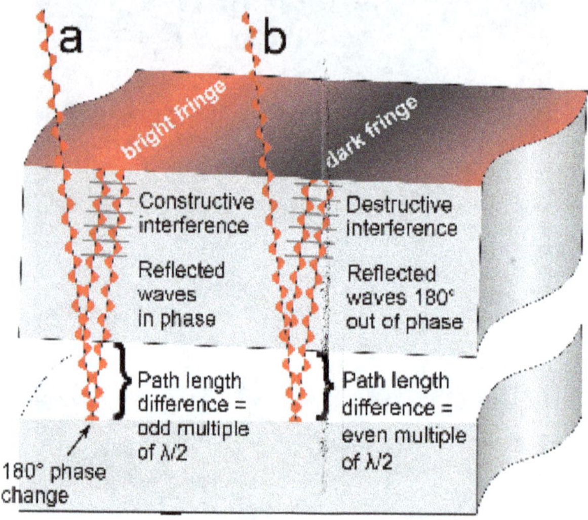

Creation of interference fringes by an optical flat on a reflective surface. Light rays from a monochromatic source pass through the glass and reflect off both the bottom surface of the flat and the supporting surface. The tiny gap between the surfaces means the two reflected rays have different path lengths. In addition the ray reflected from the bottom plate undergoes a 180° phase reversal. As a result, at locations ***(a)*** where the path difference is an odd multiple of λ/2, the waves reinforce. At locations ***(b)*** where the path difference is an even multiple of λ/2 the waves cancel. Since the gap between the surfaces varies slightly in width at different points, a series of alternating bright and dark bands, *interference fringes*, are seen.

Because the frequency of light waves ($\sim 10^{14}$ Hz) is too high to be detected by currently available detectors, it is possible to observe only the intensity of an optical interference pattern. The intensity of the light at a given point is proportional to the square of the average amplitude of the wave. This can be expressed mathematically as follows. The displacement of the two waves at a point r is:

$$U_1(\mathbf{r},t) = A_1(\mathbf{r})e^{i[\varphi_1(\mathbf{r})-\omega t]}$$

$$U_2(\mathbf{r},t) = A_2(\mathbf{r})e^{i[\varphi_2(\mathbf{r})-\omega t]}$$

where A represents the magnitude of the displacement, φ represents the phase and ω represents the angular frequency.

The displacement of the summed waves is

$$U(\mathbf{r},t) = A_1(\mathbf{r})e^{i[\varphi_1(\mathbf{r})-\omega t]} + A_2(\mathbf{r})e^{i[\varphi_2(\mathbf{r})-\omega t]}$$

The intensity of the light at r is given by

$$I(\mathbf{r}) = \int U(\mathbf{r},t)U^*(\mathbf{r},t)dt \propto A_1^2(\mathbf{r}) + A_2^2(\mathbf{r}) + 2A_1(\mathbf{r})A_2(\mathbf{r})\cos[\varphi_1(\mathbf{r}) - \varphi_2(\mathbf{r})]$$

This can be expressed in terms of the intensities of the individual waves as

$$I(\mathbf{r}) = I_1(\mathbf{r}) + I_2(\mathbf{r}) + 2\sqrt{I_1(\mathbf{r})I_2(\mathbf{r})}\cos[\varphi_1(\mathbf{r}) - \varphi_2(\mathbf{r})]$$

Thus, the interference pattern maps out the difference in phase between the two waves, with maxima occurring when the phase difference is a multiple of 2π. If the two beams are of equal intensity, the maxima are four times as bright as the individual beams, and the minima have zero intensity.

The two waves must have the same polarization to give rise to interference fringes since it is not possible for waves of different polarizations to cancel one another out or add together. Instead, when waves of different polarization are added together, they give rise to a wave of a different polarization state.

Light Source Requirements

The discussion above assumes that the waves which interfere with one another are monochromatic, i.e. have a single frequency—this requires that they are infinite in time. This is not, however, either practical or necessary. Two identical waves of finite duration whose frequency is fixed over that period will give rise to an interference pattern while they overlap. Two identical waves which consist of a narrow spectrum of frequency waves of finite duration, will give a series of fringe patterns of slightly differing spacings, and provided the spread of spacings is significantly less than the average fringe spacing, a fringe pattern will again be observed during the time when the two waves overlap.

Conventional light sources emit waves of differing frequencies and at different times from different points in the source. If the light is split into two waves and then re-combined, each individual light wave may generate an interference pattern with its other half, but the individual fringe patterns generated will have different phases and spacings, and normally no overall fringe pattern will be observable. However, single-element light sources, such as sodium- or mercury-vapor lamps have emission lines with quite narrow frequency spectra. When these are spatially and colour filtered, and then split into two waves, they can be superimposed to generate interference fringes. All interferometry prior to the invention of the laser was done using such sources and had a wide range of successful applications.

A laser beam generally approximates much more closely to a monochromatic source, and it is much more straightforward to generate interference fringes using a laser. The ease with which interference fringes can be observed with a laser beam can sometimes cause problems in that stray reflections may give spurious interference fringes which can result in errors.

Normally, a single laser beam is used in interferometry, though interference has been observed using two independent lasers whose frequencies were sufficiently matched to satisfy the phase requirements.

White light interference in a soap bubble

It is also possible to observe interference fringes using white light. A white light fringe pattern can be considered to be made up of a 'spectrum' of fringe patterns each of slightly different spacing. If all the fringe patterns are in phase in the centre, then the fringes will increase in size as the wavelength decreases and the summed intensity will show three to four fringes of varying colour. Young describes this very elegantly in his discussion of two slit interference. Some fine examples of white light fringes can be seen here. Since white light fringes are obtained only when the two waves have travelled equal distances from the light source, they can be very useful in interferometry, as they allow the zero path difference fringe to be identified.

Optical Arrangements

To generate interference fringes, light from the source has to be divided into two waves which have then to be re-combined. Traditionally, interferometers have been classified as either amplitude-division or wavefront-division systems.

In an amplitude-division system, a beam splitter is used to divide the light into two beams travelling in different directions, which are then superimposed to produce the interference pattern. The Michelson interferometer and the Mach-Zehnder interferometer are examples of amplitude-division systems.

In wavefront-division systems, the wave is divided in space—examples are Young's double slit interferometer and Lloyd's mirror.

Interference can also be seen in everyday phenomena such as iridescence and structural coloration. For example, the colours seen in a soap bubble arise from interference of light reflecting off the front and back surfaces of the thin soap film. Depending on the thickness of the film, different colours interfere constructively and destructively.

Applications

Optical Interferometry

Interferometry has played an important role in the advancement of physics, and also has a wide range of applications in physical and engineering measurement.

Thomas Young's double slit interferometer in 1803 demonstrated interference fringes when two small holes were illuminated by light from another small hole which was illuminated by sunlight.

Young was able to estimate the wavelength of different colours in the spectrum from the spacing of the fringes. The experiment played a major role in the general acceptance of the wave theory of light. In quantum mechanics, this experiment is considered to demonstrate the inseparability of the wave and particle natures of light and other quantum particles (wave–particle duality). Richard Feynman was fond of saying that all of quantum mechanics can be gleaned from carefully thinking through the implications of this single experiment.

The results of the Michelson–Morley experiment are generally considered to be the first strong evidence against the theory of a luminiferous aether and in favor of special relativity.

Interferometry has been used in defining and calibrating length standards. When the metre was defined as the distance between two marks on a platinum-iridium bar, Michelson and Benoît used interferometry to measure the wavelength of the red cadmium line in the new standard, and also showed that it could be used as a length standard. Sixty years later, in 1960, the metre in the new SI system was defined to be equal to 1,650,763.73 wavelengths of the orange-red emission line in the electromagnetic spectrum of the krypton-86 atom in a vacuum. This definition was replaced in 1983 by defining the metre as the distance travelled by light in vacuum during a specific time interval. Interferometry is still fundamental in establishing the calibration chain in length measurement.

Interferometry is used in the calibration of slip gauges (called gauge blocks in the US) and in coordinate-measuring machines. It is also used in the testing of optical components.

Radio Interferometry

The Very Large Array, an interferometric array formed from many smaller telescopes, like many larger radio telescopes.

In 1946, a technique called astronomical interferometry was developed. Astronomical radio interferometers usually consist either of arrays of parabolic dishes or two-dimensional arrays of omni-directional antennas. All of the telescopes in the array are widely separated and are usually connected together using coaxial cable, waveguide, optical fiber, or other type of transmission line. Interferometry increases the total signal collected, but its primary purpose is to vastly increase the resolution through a process called Aperture synthesis. This technique works by superposing (interfering) the signal waves from the different telescopes on the principle that waves that coincide with the same phase will add to each other while two waves that have opposite phases will cancel each other out. This creates a combined telescope that is equivalent in resolution (though not in sensitivity) to a single antenna whose diameter is equal to the spacing of the antennas furthest apart in the array.

Acoustic Interferometry

An acoustic interferometer is an instrument for measuring the physical characteristics of sound wave in a gas or liquid. It may be used to measure velocity, wavelength, absorption, or impedance. A vibrating crystal creates the ultrasonic waves that are radiated into the medium. The waves strike a reflector placed parallel to the crystal. The waves are then reflected back to the source and measured.

Quantum Interference

If a system is in state ψ, its wavefunction is described in Dirac or bra–ket notation as:

$$|\psi\rangle = \sum_i |i\rangle\langle i|\psi\rangle = \sum_i |i\rangle\psi_i$$

where the $|i\rangle$s specify the different quantum "alternatives" available (technically, they form an eigenvector basis) and the ψ_i are the probability amplitude coefficients, which are complex numbers.

The probability of observing the system making a transition or quantum leap from state ψ to a new state φ is the square of the modulus of the scalar or inner product of the two states:

$$\text{prob}(\psi \Rightarrow \varphi) = |\langle\psi|\varphi\rangle|^2 = |\sum_i \psi_i^* \varphi_i|^2$$

$$= \sum_{ij} \psi_i^* \psi_j \varphi_j^* \varphi_i = \sum_i |\psi_i|^2 |\varphi_i|^2 + \sum_{ij;i\neq j} \psi_i^* \psi_j \varphi_j^* \varphi_i$$

where $\psi_i = \langle i|\psi\rangle$ (as defined above) and similarly $\varphi_i = \langle i|\varphi\rangle$ are the coefficients of the final state of the system. * is the complex conjugate so that $\psi_i^* = \langle\psi|i\rangle$, etc.

Now let's consider the situation classically and imagine that the system transited from $|\psi\rangle$ to $|\varphi\rangle$ via an intermediate state $|i\rangle$. Then we would *classically* expect the probability of the two-step transition to be the sum of all the possible intermediate steps. So we would have

$$\text{prob}(\psi \Rightarrow \varphi) = \sum_i \text{prob}(\psi \Rightarrow i \Rightarrow \varphi)$$

$$= \sum_i |\langle\psi|i\rangle|^2 |\langle i|\varphi\rangle|^2 = \sum_i |\psi_i|^2 |\varphi_i|^2$$

,

The classical and quantum derivations for the transition probability differ by the presence, in the quantum case, of the extra terms $\sum_{ij;i\neq j} \psi_i^* \psi_j \varphi_j^* \varphi_i$; these extra quantum terms represent *interference* between the different $i \neq j$ intermediate "alternatives". These are consequently known as the quantum interference terms, or cross terms. This is a purely quantum effect and is a consequence of the non-additivity of the probabilities of quantum alternatives.

The interference terms vanish, via the mechanism of quantum decoherence, if the intermediate state $|i\rangle$ is measured or coupled with the environment.

Consider a situation where we superpose two waves. Naively, we would expect the intensity (energy density or flux) of the resultant to be the sum of the individual intensities. For example, a room becomes twice as bright if we switch on two lamps instead of one. This actually does not always hold. A wave, unlike the intensity, can have a negative value. If we add two waves

whose values have opposite signs at the same point, the total intensity is less than the intensities of the individual waves. This is an example of a phenomena referred to as interference.

Young's Double Slit Experiment.

We begin our discussion of interference with a situation shown in Figure. Light from a distant point source is incident on a screen with two thin slits. The separation between the two slits is d. We are interested in the image of the two slits on a screen which is at a large distance from the slits. Note that the point source is aligned with the center of the slits as shown in Figure. Let us calculate the intensity at a point P located at an angle θ on the screen.

The radiation from the point source is well described by a plane wave by the time the radiation reaches the slits. The two slits lie on the same wavefront of this plane wave, thus the electric field oscillates with the same phase at both

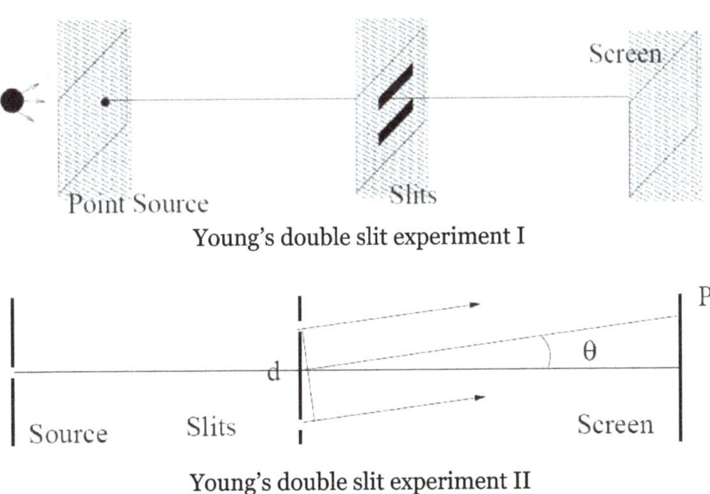

Young's double slit experiment I

Young's double slit experiment II

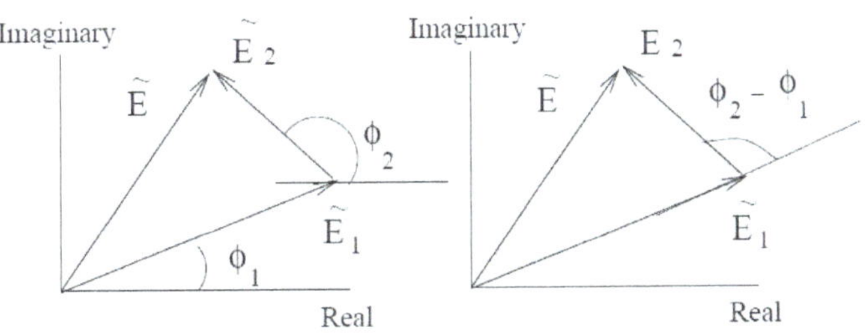

Summation of two phasors

the slits. If $\tilde{E}_1(t)$ *and* $\tilde{E}_2(t)$ be the contributions from slits 1 and 2 to the radiation at the point P on the screen, the total electric field will be

$$\tilde{E}(t) = \tilde{E}_1(t) + \tilde{E}_2(t)$$

Both waves originate from the same source and they have the same frequency. We can thus express the electric fields as $\tilde{E}_1(t) = \tilde{E}_1 e^{i\omega t}, \tilde{E}_2(t) = \tilde{E}_2 e^{i\omega t}$ *and* $\tilde{E}(t) = \tilde{E} e^{i\omega t}$. We then have a relations between the amplitudes

$$\tilde{E} = \tilde{E}_1 + \tilde{E}_2$$

It is often convenient to represent this addition of complex amplitudes graphically as shown in Figure. Each complex amplitude can be represented by a vector in the complex plane, such a vector is called a phasor. The sum is now a vector sum of the phasors.

The intensity of the wave is

$$I = \langle E(t)E(t) \rangle = \frac{1}{2}\tilde{E}\tilde{E}^*$$

where we have dropped the constant of proportionatily in this relation. It is clear that the square of the length of the resultant phasor gives the intensity. Geometrically, the resultant intensity I is the square of the vector sum of two vectors of length $\sqrt{I_1}$ *and* $\sqrt{I_2}$ with angle $\phi_2 - \phi_1$ between them as shown in Figure. Consequently, the resulting intensity is

$$I = I_1 + I_2 + 2\sqrt{I_1 I_2}\cos(\phi_2 - \phi_1).$$

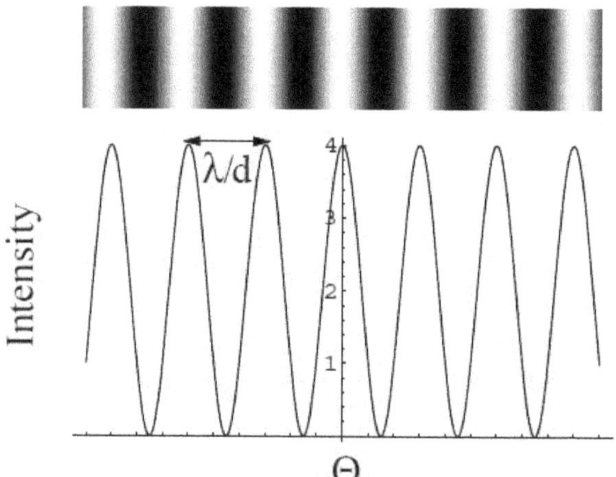

Young double slit interference fringes with intensity profile

Calculating the intensity algebraically, we see that it is

$$I = \frac{1}{2}\left[\tilde{E}_1\tilde{E}_1^* + \tilde{E}_2\tilde{E}_2^* + \tilde{E}_1\tilde{E}_2^* + \tilde{E}_1^*\tilde{E}_2\right]$$

$$= I_1 + I_2 + \frac{1}{2} E_1 E_2 \left[e^{i(\phi_1 - \phi_2)} + e^{i(\phi_2 - \phi_1)} \right]$$

$$I = I_1 + I_2 + 2\sqrt{I_1 I_2} \cos(\phi_2 - \phi_1)$$

The intensity is maximum when the two waves have the same phase

$$\mathrm{I} = \mathrm{I}_1 + \mathrm{I}_2 + 2\sqrt{I_1 I_2}$$

and it is minimum when $\phi_2 - \phi_1 = \pi$ i.e the two waves are exactly out of phase

$$I = I_1 + I_2 - 2\sqrt{I_1 I_2}\,.$$

The intensity is the sum of the two intensities when the two waves are $\pi/2$ out of phase.

In the Young's double slit experiment the waves from the two slits arrive at P with a time delay because the two waves have to traverse different paths. The resulting phase difference is

$$\phi_1 - \phi_2 = 2\pi \frac{d \sin\theta}{\lambda}.$$

If the two slits are of the same size and are equidistant from the the original source, then $I_1 = I_2$ and the resultant intensity,

$$I(\theta) = 2I_1 \left[1 + \cos\left(\frac{2\pi d \sin\theta}{\lambda} \right) \right]$$

$$= 4I_1 \cos^2 \left(\frac{\pi d \sin\theta}{\lambda} \right)$$

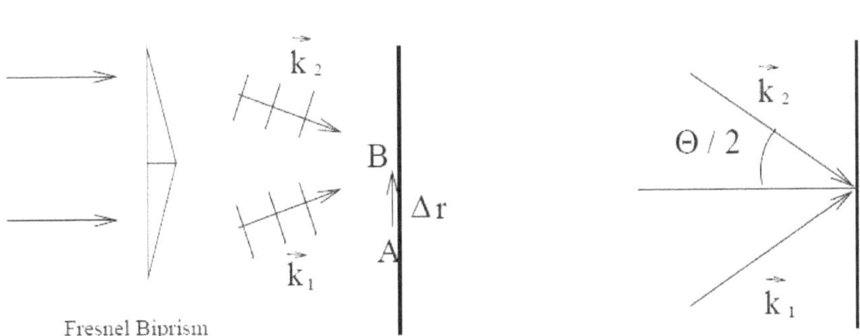

Fresnel biprism to realise the Young's double slit

For small θ we have

$$I(\theta) = 2I_1 \left[1 + \cos\left(\frac{2\pi d\theta}{\lambda} \right) \right]$$

There will be a pattern of bright and dark lines, referred to as fringes, that will be seen on the screen as in Figure. The fringes are straight lines parallel to the slits, and the spacing between two successive bright fringes is λ / d radians.

A different Method of Analysis.

A Fresnel biprism is constructed by joining to identical thin prisms as shown in Figure. Consider a plane wave from a distant point source incident on the Fresnel biprism. The part of the wave that passes through the upper half of the biprism propagates in a slightly different direction from the part that passes through the lower half of the biprism. The light emanating from the biprism is equivalent to that from two exactly identical sources, the sources being located far away and there being a small separation between the sources. The Fresnel biprism provides a method for implementing the Young's double slit experiment.

The two waves emanating from the biprism will be coplanar and in different directions with wave vectors $\vec{k}_1$ *and* $\vec{k}_2$ as shown in Figure. We are interested in the intensity distribution on the screen shown in the figure. Let A be a point where both waves arrive at the same phase.ie $\phi(A)$ *ie.* $\tilde{E}_1 = \tilde{E}_2 = Ee^{i\phi(A)}$. The intensity at this point will be a maximum. Next consider a point B at a displacement $\Delta\vec{r}$ from the point A. The phase of the two waves are different at this point. The phase of the first wave at the point B is given by

$$\phi_1(B) = \phi(A) - \vec{k}_1 \cdot \Delta\vec{r}$$

and far the second wave

$$\phi_2(B) = \phi(A) - \vec{k}_1 \cdot \Delta\vec{r}$$

The phase difference is

$$\phi_2 - \phi_1 = \left(\vec{k}_1 - \vec{k}_2\right) \cdot \Delta\vec{r}$$

Using eq., the intensity pattern on the screen is given by

$$I(\Delta\vec{r}) = I_1 + I_2 + 2\sqrt{I_1 I_2}\,\cos[\left(\vec{k}_1 - \vec{k}_2\right) \cdot \Delta\vec{r}]$$

where I_1 and I_2 are the intensities of the waves from the upper and lower half of the biprism respectively. Assuming that the wave vectors make a small angle $\theta / 2 \ll 1$ to the horizontal we have

$$\vec{k}_1 = k\left[\hat{i} + \frac{\theta}{2}\hat{j}\right] \text{ and } \vec{k}_2 = k\left[\hat{i} - \frac{\theta}{2}\hat{j}\right]$$

where θ is the angle between the two waves. Using this and assuming that $I_1 = I_2$ we have

$$I(\Delta\vec{r}) = 2I_1\left[1 + \cos\left(\frac{2\pi\theta\,\Delta y}{\lambda}\right)\right].$$

There will be straight line fringes on the screen, these fringes are perpendicular to the y axis and have a fringe spacing $\Delta y = \lambda / \theta$. The analysis presented here is another way of analysing the Young's double slit experiment. It is left to the reader to verify that eq. are equivalent. Like Fresnel biprism one can also realise double slit experiment with 'Fresnel mirrors'. Here one uses two plane mirrors and one of the mirrors is tilted slightly $(\theta < 1^0)$ and glued with the other as shown in Figure.

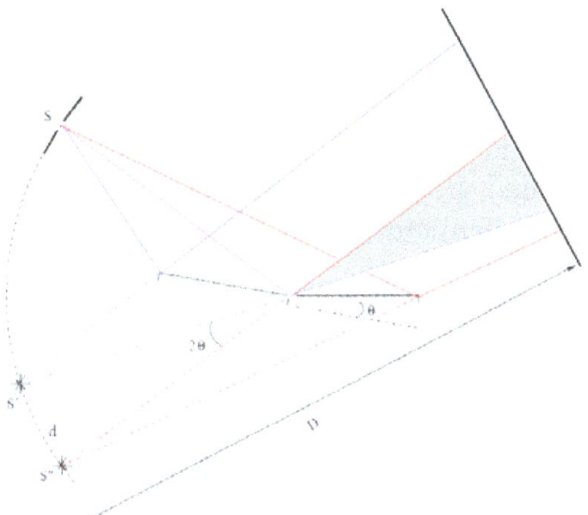

Fresnel mirrors

Michelson Interferometer

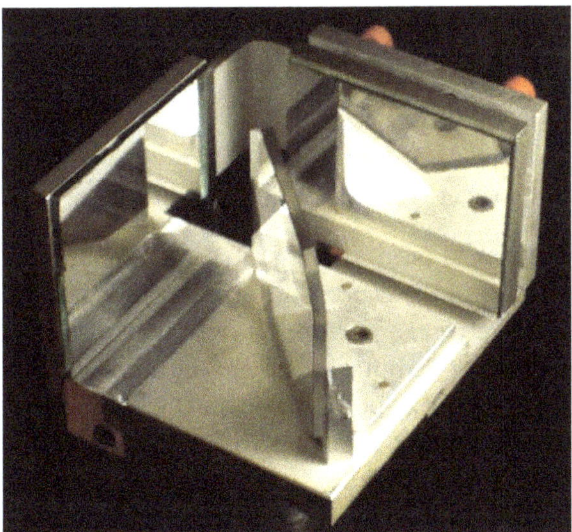

A basic Michelson interferometer, not including the optical source and detector.

The Michelson interferometer is a common configuration for optical interferometry and was invented by Albert Abraham Michelson. Using a beamsplitter, a light source is split into two arms. Each of those is reflected back toward the beamsplitter which then combines their amplitudes interferometrically. The resulting interference pattern that is not directed back toward the source is typically directed to some type of photoelectric detector or camera. Depending on the interferometer's particular application, the two paths may be of different lengths or include optical materials or components under test.

The Michelson interferometer (among other interferometer configurations) is employed in many scientific experiments and became well known for its use by Albert Michelson and Edward Morley in the famous Michelson-Morley experiment (1887) in a configuration which would have detected the earth's motion through the supposed luminiferous aether that most physicists at the time believed was the medium in which light waves propagated. The null result of that experiment essentially disproved the existence of such an aether, leading eventually to the special theory of relativity and the revolution in physics at the beginning of the twentieth century. In 2016, another application of the Michelson interferometer, LIGO, made the first direct detection of gravitational waves. That observation confirmed an important prediction of general relativity, validating the theory's prediction of space-time distortion in the context of large scale cosmic events (known as strong field tests).

Configuration

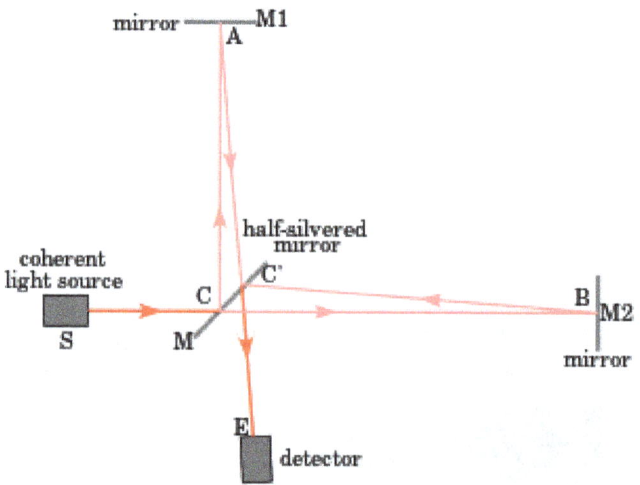

Path of light in Michelson interferometer.

A Michelson interferometer consists minimally of mirrors M_1 & M_2 and a beam splitter *M*. In Figure, a source *S* emits light that hits the beam splitter (in this case, a plate beamsplitter) surface *M* at point *C*. *M* is partially reflective, so part of the light is transmitted through to point *B* while some is reflected in the direction of *A*. Both beams recombine at point *C'* to produce an interference pattern incident on the detector at point *E* (or on the retina of a person's eye). If there is a slight angle between the two returning beams, for instance, then an imaging detector will record a sinusoidal *fringe pattern* as shown in Figure b. If there is perfect spatial alignment between the returning beams, then there will not be any such pattern but rather a constant intensity over the beam dependent on the differential pathlength; this is difficult, requiring very precise control of the beam paths.

Figure shows use of a coherent (laser) source. Narrowband spectral light from a discharge or even white light can also be used, however to obtain significant interference contrast it is required that the differential pathlength is reduced below the coherence length of the light source. That can be only micrometers for white light, as discussed below.

If a lossless beamsplitter is employed, then one can show that optical energy is conserved. At every point on the interference pattern, the power that is *not* directed to the detector at *E* is rather present in a beam (not shown) returning in the direction of the source.

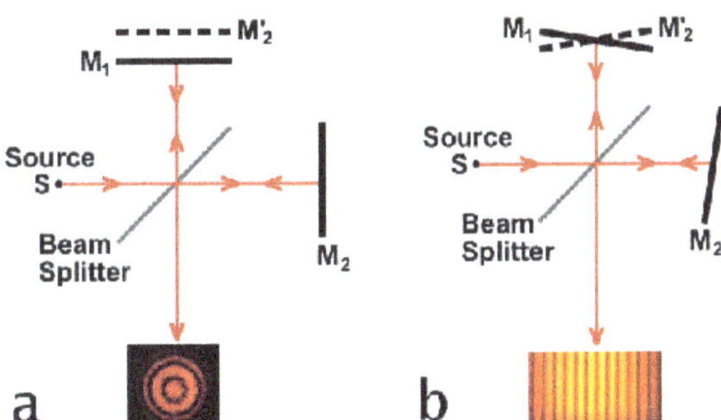

Formation of fringes in a Michelson interferometer

As shown in Figure a and b, the observer has a direct view of mirror M_1 seen through the beam splitter, and sees a reflected image M'_2 of mirror M_2. The fringes can be interpreted as the result of interference between light coming from the two virtual images S'_1 and S'_2 of the original source *S*. The characteristics of the interference pattern depend on the nature of the light source and the precise orientation of the mirrors and beam splitter. In Figure a, the optical elements are oriented so that S'_1 and S'_2 are in line with the observer, and the resulting interference pattern consists of circles centered on the normal to M_1 and M'_2 (fringes of equal inclination). If, as in Figure b, M_1 and M'_2 are tilted with respect to each other, the interference fringes will generally take the shape of conic sections (hyperbolas), but if M_1 and M'_2 overlap, the fringes near the axis will be straight, parallel, and equally spaced (fringes of equal thickness). If S is an extended source rather than a point source as illustrated, the fringes of Figure a must be observed with a telescope set at infinity, while the fringes of Figure b will be localized on the mirrors.

Source Bandwidth

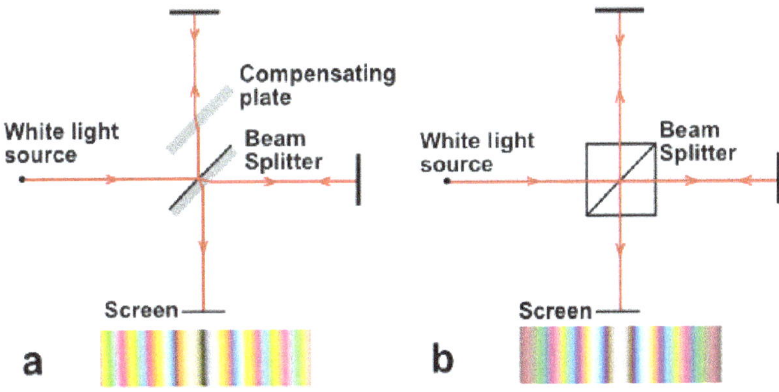

Michelson interferometers using a white light source

White light has a tiny coherence length and is difficult to use in a Michelson (or Mach-Zehnder) interferometer. Even a narrowband (or "quasi-monochromatic") spectral source requires careful attention to issues of chromatic dispersion when used to illuminate an interferometer. The two optical paths must be practically equal for all wavelengths present in the source. This requirement can be met if both light paths cross an equal thickness of glass of the same dispersion. In Figure a, the horizontal beam crosses the beam splitter three times, while the vertical beam crosses the beam splitter once. To equalize the dispersion, a so-called compensating plate identical to the substrate of the beam splitter may be inserted into the path of the vertical beam. In Figure b, we see using a cube beam splitter already equalizes the pathlengths in glass. The requirement for dispersion equalization is eliminated by using extremely narrowband light from a laser.

The extent of the fringes depends on the coherence length of the source. In Figure b, the yellow sodium light used for the fringe illustration consists of a pair of closely spaced lines, D_1 and D_2, implying that the interference pattern will blur after several hundred fringes. Single longitudinal mode lasers are highly coherent and can produce high contrast interference with differential pathlengths of millions or even billions of wavelengths. On the other hand, using white (broadband) light, the central fringe is sharp, but away from the central fringe the fringes are colored and rapidly become indistinct to the eye.

Early experimentalists attempting to detect the earth's velocity relative to the supposed luminiferous aether, such as Michelson and Morley (1887) and Miller (1933), used quasi-monochromatic light only for initial alignment and coarse path equalization of the interferometer. Thereafter they switched to white (broadband) light, since using white light interferometry they could measure the point of *absolute phase* equalization (rather than phase modulo 2π), thus setting the two arms' pathlengths equal. More importantly, in a white light interferometer, any subsequent "fringe jump" (differential pathlength shift of one wavelength) would always be detected.

Applications

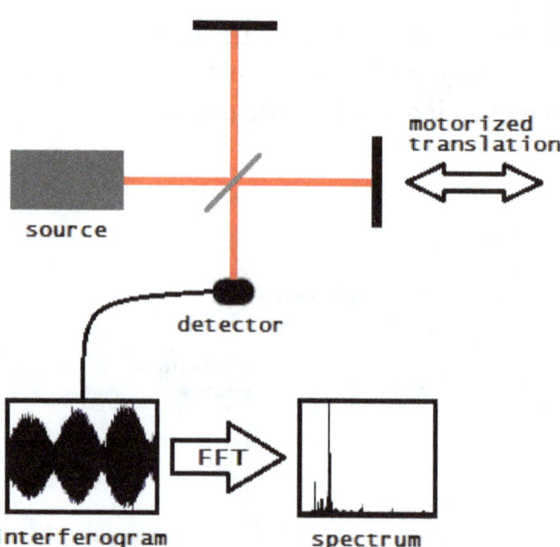

Fourier transform spectroscopy.

The Michelson interferometer configuration is used in a number of different applications.

Fourier Transform Spectrometer

Figure illustrates the operation of a Fourier transform spectrometer, which is essentially a Michelson interferometer with one mirror movable. (A practical Fourier transform spectrometer would substitute corner cube reflectors for the flat mirrors of the conventional Michelson interferometer, but for simplicity, the illustration does not show this.) An interferogram is generated by making measurements of the signal at many discrete positions of the moving mirror. A Fourier transform converts the interferogram into an actual spectrum. Fourier transform spectrometers can offer significant advantages over dispersive (*i.e.* grating and prism) spectrometers under certain conditions. (1) The Michelson interferometer's detector in effect monitors all wavelengths simultaneously throughout the entire measurement. When using a noisy detector, such as at infrared wavelengths, this offers an increase in signal to noise ratio while using only a single detector element; (2) the interferometer does not require a limited aperture as do grating or prism spectrometers, which require the incoming light to pass through a narrow slit in order to achieve high spectral resolution. This is an advantage when the incoming light is not of a single spatial mode. For more information.

Twyman-Green Interferometer

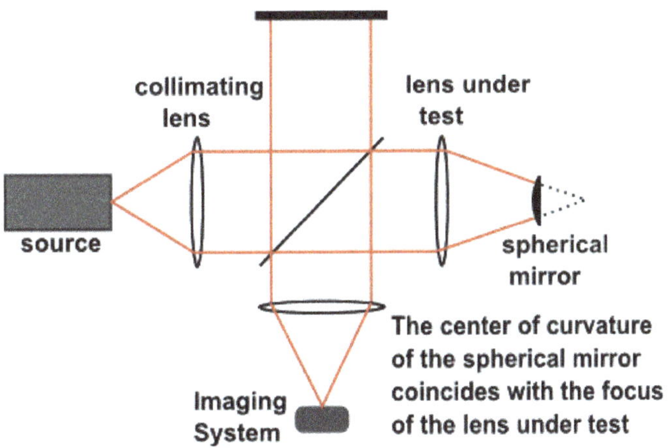

Twyman-Green Interferometer.

The Twyman-Green interferometer is a variation of the Michelson interferometer used to test small optical components, invented and patented by Twyman and Green in 1916. The basic characteristics distinguishing it from the Michelson configuration are the use of a monochromatic point light source and a collimator. It is interesting to note that Michelson (1918) criticized the Twyman-Green configuration as being unsuitable for the testing of large optical components, since the available light sources had limited coherence length. Michelson pointed out that constraints on geometry forced by the limited coherence length required the use of a reference mirror of equal size to the test mirror, making the Twyman-Green impractical for many purposes. Decades later, the advent of laser light sources answered Michelson's objections.

The use of a figured reference mirror in one arm allows the Twyman-Green interferometer to be used for testing various forms of optical component, such as lenses or telescope mirrors. Figure illustrates a Twyman-Green interferometer set up to test a lens. A point source of monochromatic light is expanded by a diverging lens (not shown), then is collimated into a parallel beam. A convex spherical mirror is positioned so that its center of curvature coincides with the focus of the lens being tested. The emergent beam is recorded by an imaging system for analysis.

Laser Unequal Path Interferometer

The "LUPI" is a Twyman-Green interferometer that uses a coherent laser light source. The high coherence length of a laser allows unequal path lengths in the test and reference arms and permits economical use of the Twyman-Green configuration in testing large optical components.

Step-phase Interferometer

This is a Michelson interferometer in which the mirror in one arm is replaced with a Gires–Tournois etalon. The highly dispersed wave reflected by the Gires–Tournois etalon interferes with the original wave as reflected by the other mirror. Because the phase change from the Gires–Tournois etalon is an almost step-like function of wavelength, the resulting interferometer has special characteristics. It has an application in fiber-optic communications as an optical interleaver.

Both mirrors in a Michelson interferometer can be replaced with Gires–Tournois etalons. The step-like relation of phase to wavelength is thereby more pronounced, and this can be used to construct an asymmetric optical interleaver.

Stellar Measurements

The Michelson stellar interferometer is used for measuring the diameter of stars.

Gravitational Wave Detection

Michelson interferometry is one leading method for the direct detection of gravitational waves. This involves detecting tiny strains in space itself, affecting two long arms of the interferometer unequally, due to a strong passing gravitational wave. In 2015 the first detection of gravitational waves was accomplished using the LIGO instrument, a Michelson interferometer with 4 km arms. This was the first experimental validation of gravitational waves, predicted by Albert Einstein's General Theory of Relativity. An even larger Michelson interferometer in space, to achieve greater sensitivity, is in the planning stages.

Miscellaneous Applications

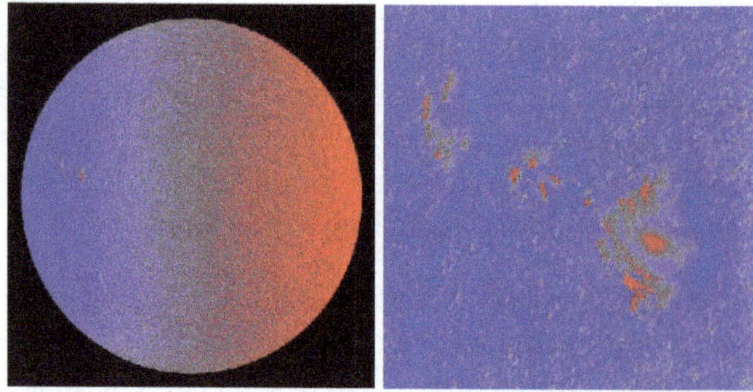

Helioseismic Magnetic Imager (HMI) dopplergram showing the velocity of gas flows on the solar surface. Red indicates motion away from the observer, and blue indicates motion towards the observer.

Figure illustrates use of a Michelson interferometer as a tunable narrow band filter to create dopplergrams of the Sun's surface. When used as a tunable narrow band filter, Michelson interferometers exhibit a number of advantages and disadvantages when compared with competing technologies such as Fabry–Pérot interferometers or Lyot filters. Michelson interferometers have the largest field of view for a specified wavelength, and are relatively simple in operation, since tuning is via mechanical rotation of waveplates rather than via high voltage control of piezoelectric crystals or lithium niobate optical modulators as used in a Fabry–Pérot system. Compared with Lyot filters, which use birefringent elements, Michelson interferometers have a relatively low temperature sensitivity. On the negative side, Michelson interferometers have a relatively restricted wavelength range, and require use of prefilters which restrict transmittance. The reliability of Michelson interferometers has tended to favor their use in space applications, while the broad wavelength range and overall simplicity of Fabry–Pérot interferometers has favored their use in ground-based systems.

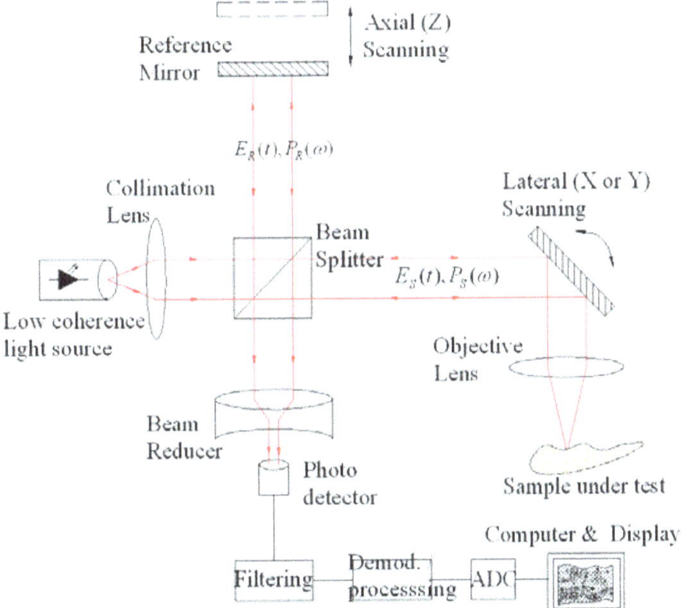

Typical optical setup of single point OCT

Another application of the Michelson Interferometer is in optical coherence tomography (OCT), a medical imaging technique using low-coherence interferometry to provide tomographic visualization of internal tissue microstructures. As seen in Figure, the core of a typical OCT system is a Michelson interferometer. One interferometer arm is focused onto the tissue sample and scans the sample in an X-Y longitudinal raster pattern. The other interferometer arm is bounced off a reference mirror. Reflected light from the tissue sample is combined with reflected light from the reference. Because of the low coherence of the light source, interferometric signal is observed only over a limited depth of sample. X-Y scanning therefore records one thin optical slice of the sample at a time. By performing multiple scans, moving the reference mirror between each scan, an entire three-dimensional image of the tissue can be reconstructed. Recent advances have striven to combine the nanometer phase retrieval of coherent interferometry with the ranging capability of low-coherence interferometry.

Another application is a sort of delay line interferometer that converts phase modulation into amplitude modulation in DWDM networks.

Atmospheric and Space Applications

The Michelson Interferometer has played an important role in studies of the upper atmosphere, revealing temperatures and winds, employing both space-borne, and ground-based instruments, by measuring the Doppler widths and shifts in the spectra of airglow and aurora. For example, the Wind Imaging Interferometer, WINDII, on the Upper Atmosphere Research Satellite, UARS, (launched on September 12, 1991) measured the global wind and temperature patterns from 80 to 300 km by using the visible airglow emission from these altitudes as a target and employing optical Doppler interferometry to measure the small wavelength shifts of the narrow atomic and molecular airglow emission lines induced by the bulk velocity of the atmosphere carrying the emitting species. The instrument was an all-glass field-widened achromatically and thermally compensated phase-stepping Michelson interferometer, along with a bare CCD detector that imaged the airglow limb through the interferometer. A sequence of phase-stepped images was processed to derive the wind velocity for two orthogonal view directions, yielding the horizontal wind vector.

The principle of using a polarizing Michelson Interferometer as a narrow band filter was first described by Evans who developed a birefringent photometer where the incoming light is split into two orthogonally polarized components by a polarizing beam splitter, sandwiched between two halves of a Michelson cube. This led to the first polarizing wide-field Michelson interferometer described by Title and Ramsey which was used for solar observations; and led to the development of a refined instrument applied to measurements of oscillations in the sun's atmosphere, employing a network of observatories around the Earth known as the Global Oscillations Network Group (GONG).

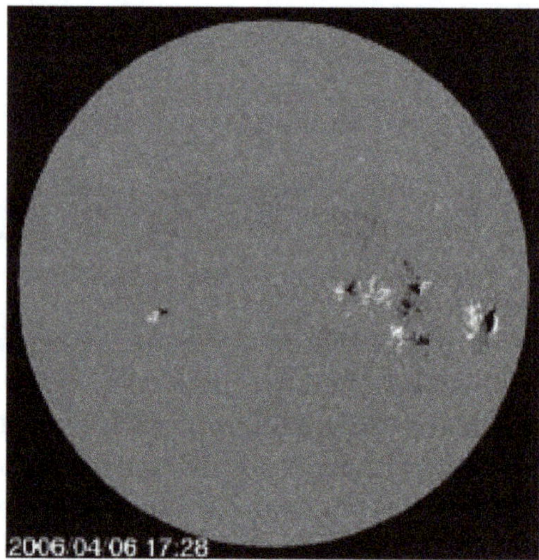

Magnetogram (magnetic image) of the Sun showing magnetically intense areas (active regions) in black and white, as imaged by the Helioseismic and Magnetic Imager (HMI) on the Solar Dynamics Observatory

The Polarizing Atmospheric Michelson Interferometer, PAMI, developed by Bird et al., and discussed in *Spectral Imaging of the Atmosphere*, combines the polarization tuning technique of Title and Ramsey with the Shepherd *et al.* technique of deriving winds and temperatures from emission rate measurements at sequential path differences, but the scanning system used by PAMI is much simpler than the moving mirror systems in that it has no internal moving parts, instead scanning with a polarizer external to the interferometer. The PAMI was demonstrated in an observation campaign where its performance was compared to a Fabry–Pérot spectrometer, and employed to measure E-region winds.

More recently, the Helioseismic and Magnetic Imager (HMI), on the Solar Dynamics Observatory, employs two Michelson Interferometers with a polarizer and other tunable elements, to study solar variability and to characterize the Sun's interior along with the various components of magnetic activity. HMI takes high-resolution measurements of the longitudinal and vector magnetic field over the entire visible disk thus extending the capabilities of its predecessor, the SOHO's MDI instrument. HMI produces data to determine the interior sources and mechanisms of solar variability and how the physical processes inside the Sun are related to surface magnetic field and activity. It also produces data to enable estimates of the coronal magnetic field for studies of variability in the extended solar atmosphere. HMI observations will help establish the relationships between the internal dynamics and magnetic activity in order to understand solar variability and its effects.

In one example of the use of the MDI, Stanford scientists reported the detection of several sunspot regions in the deep interior of the Sun, 1–2 days before they appeared on the solar disc. The detection of sunspots in the solar interior may thus provide valuable warnings about upcoming surface magnetic activity which could be used to improve and extend the predictions of space weather forecasts.

Figure shows a typical Michelson interferometer setup. A ground glass plate G is illuminated by a light source. The ground glass plate has the property that it scatters the incident light into all directions. Each point on the ground glass plate acts like a source that emits light in all directions.

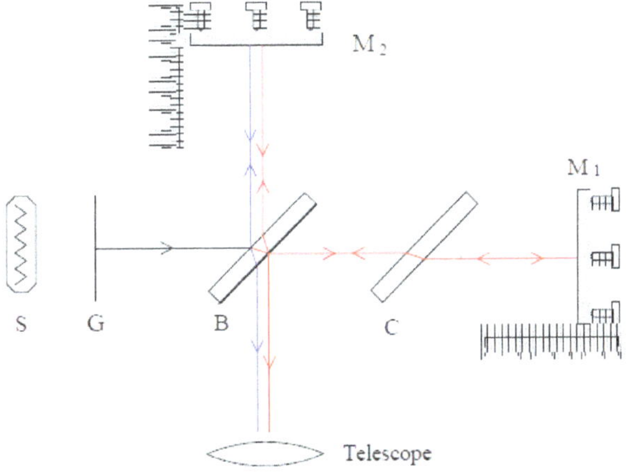

Michelson Interferometer

The light scattered forward by G is incident on a beam splitter B which is at 45°. The beam splitter is essentially a glass slab with the lower surface semi-silvered to increase its reflectivity. It splits the incident wave into two parts $\tilde{E}_1$ and $\tilde{E}_2$, one which is transmitted ($\tilde{E}_1$) and another $(\tilde{E}_2)$ which is reflected. The two beams have nearly the same intensity. The transmitted wave $\tilde{E}_1$ is reflected back to B by a mirror M_1 and a part of it is reflected into the telescope T. The reflected wave $\tilde{E}_2$ travels in a perpendicular direction. The mirror M_2 reflects this back to B where a part of it is transmitted into T. An observer at T would see two images of G, namely G_1 and G_2 produced by the two mirrors M_1 and M_2 respectively. The two images are at a separation $2d$ where d is the difference in the optical paths from B to G_1 and from B to G_2. Note that $\tilde{E}_2$ traverses the thickness of the beam splitter thrice whereas $\tilde{E}_1$ traverses the beam splitter only once. This introduces an extra optical path for $\tilde{E}_2$ even when M_1 and M_2 are at the same radiation distance from B. It is possible to compensate for this by introducing an extra displacement in M_1, but this would not serve to compensate for the extra path over a range of frequencies as the refractive index of the glass in B is frequency dependent. A compensator C, which is a glass block identical to B (without the silver coating), is introduced along the path to M_1 to compensate for this.

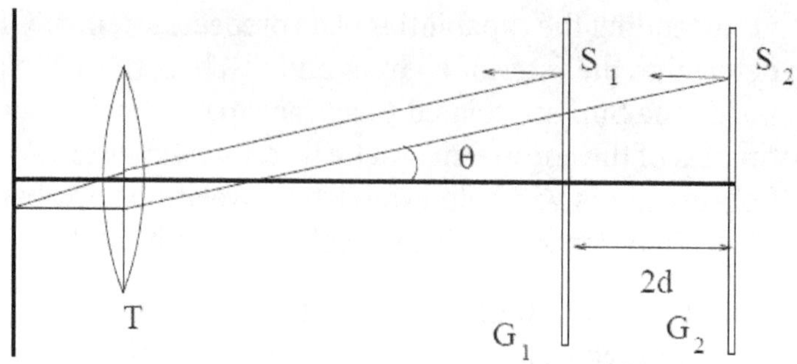

Effective set-up for Michelson Interferometer

S_1 and S_2 are the two images of the same point S on the ground glass plate. Each point on the ground glass plate acts as a source emitting radiation in all directions. Thus S_1 and S_2 are coherent sources which emit radiation in all direction. Consider the wave emitted at an angle θ as shown in Figure. The telescope focuses both waves to the same point. The resultant electric field is

$$\tilde{E} = \tilde{E}_1 + \tilde{E}_2$$

and the intensity is

$$I = I_1 + I_2 + 2\sqrt{I_1 I_2}\cos(\phi_2 - \phi_1)$$

The phase difference arises because of the path difference in the two arms of the interferometer. Further, there is an additional phase difference of π because $\tilde{E}_2$ undergoes internal reflection at B whereas $\tilde{E}_1$ undergoes external reflection. We then have

$$\phi_2 - \phi_1 = \pi + 2d\cos\theta\frac{2\pi}{\lambda}$$

So we have the condition

$$2d\cos\theta_m = m\lambda \qquad (m = 0,1,2,...)$$

for a minima or a dark fringe. Here m is the order of the fringe, and θ_m is the angle of the mth order fringe. Similarly, we have

$$2d\cos\theta_m = \left(m + \frac{1}{2}\right)\lambda \qquad (m = 0,1,2,...)$$

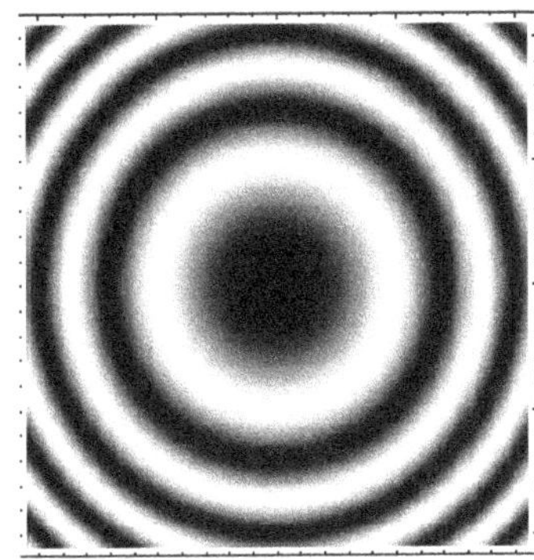

Michelson fringes

as the condition for a bight fringe. The fringes will be circular as shown in Figure. When the central fringe is dark, the order of the fringe is

$$m = \frac{2d}{\lambda}.$$

Le us follow a fringe of a fixed order, say m, as we increase d the difference in the length of the two arms. The value of cos θ_m has to decrease which implies that θ_m increases. As d is increased, new fringes appear at the center, and the existing fringes move outwards and finally move out of the field of view. For any value of d, the central fringe has the largest value of m, and the value of m decreases outwards from the center.

Considering the situation where there is a central dark fringe as shown in the left of Figure, let us estimate θ the radius of the first dark fringe. The central dark fringe satisfies the condition

$$2d = m\lambda$$

and the first dark fringe satisfies

$$2d \cos\theta = (m-1)\lambda$$

For small θ *ie.* $\theta \ll 1$ we can write eq. as

$$2d\,(1-\frac{\theta^2}{2}) = (m-1)\lambda$$

which with eq. (14.7) gives

$$\theta = \sqrt{\frac{\lambda}{d}}$$

Compare this with the Young's double slit where the fringe separation is λ / d.

The Michelson interferometer can be used to determine the wavelength of light. Consider a situation where we initially have a dark fringe at the center. This satisfies the condition given by eq. where λ, d and m are all unknown. One of the mirrors is next moved so as to increase d the difference in the lengths of the two arms of the interferometer. As the mirror is moved, the central dark fringe expands and moves out while a bright fringe appears at the center. A dark fringe reappears at the center if the mirror is moved further. The mirror is moved a distance Δd so that N new dark fringes appear at the center. Although initially d and m were unknown for the central dark fringe, it is known that finally the difference in lengths is $d+\Delta d$ and the central dark fringe is of order $N+m$ and hence it satisfies

$$2(d+\Delta d)=(m+N)\lambda$$

Subtracting eq. from this gives the wavelength of light to be

$$\lambda=\frac{2\Delta d}{N}$$

We next consider a situation where there are two very close spectral lines λ_1 and $\lambda_1+\Delta\lambda$. Each wavelength will produce its own fringe pattern. Concordance refers to the situation where the two fringe patterns coincide at the center

$$2d=m_1\lambda_1=m_2(\lambda_1+\Delta\lambda)$$

and the fringe pattern is very bright. As d is increased, m_1 and m_2 increase by different amounts with $\Delta m_2<\Delta m_1$. When $m_2=m_1-1/2$, the bright fringes of λ_1 coincide with the dark fringes of $\lambda_1+\Delta\lambda$ and vice-versa, and consequently the fringe pattern is washed away. The two set of fringes are now said to be discordant.

It is possible to measure $\Delta\lambda$ by increasing d to $d+\Delta d$ so that the two sets of fringes that are initially concordant become discordant and are finally concordant again. It is clear that if m_1 changes to $m_1+\Delta m, m_2$ changes to $m_2+\Delta m-1$ when the fringes are concordant again. We then have

$$2(d+\Delta d)=(m_1+\Delta m)\lambda_1=(m_2+\Delta m-1)(\lambda_1+\Delta\lambda)$$

which gives

$$\lambda_1=\left(\frac{2\Delta d}{\lambda_1}-1\right)\Delta\lambda$$

where on assuming that $2\Delta d/\lambda_1=m_1\gg 1$ we have

$$\partial\lambda=\frac{\lambda_1^2}{\Delta d}.$$

The Michelson interferometer finds a variety of other application. It was used by Michelson and Morley in 1887 to show that the speed of light is the same in all directions. The armlength of their interferometer was 11m. Since the Earth is moving, we would expect the speed of light to be different along the direction of the Earth's motion. Michelson and Morley established that the speed of light does not depend on the motion of the observer, providing a direct experimental basis for Einstein's Special Theory of Relativity.

Laser Interferometer Gravitational-Wave Observatory

The fringe patter in the Michelson interferometer is very sensitive to changes in the mirror positions, and it can be used to measure very small displacements of the mirrors. A Michelson interferometer whose arms are 4 km long is being used in an experiment called Laser Interferometer Gravitational- Wave Observatory (LIGO[1]) which is an ongoing effort to detect Gravitational Waves, one of the predictions of Einstein's General Theory of Relativity. Gravitational waves are disturbances in space-time that propagate at the speed of light. A gravitational wave that passes through the Michelson interferometer will produce displacements in the mirrors and these will cause changes in the fringe pattern. These displacements are predicted to be extremely small. LIGO is sensitive enough to detect displacements of the order of 10−16 cm in the mirror positions.

Two Beam interference: Newton's Rings

In a previous chapter we studied the two beam interference with Young's double slit. Realisation of Young's double slit with Fresnel biprism uses division of wavefront. Interference can also be observed where the apparatus uses division of amplitude. In Newton's rings one finds division of amplitude. The basic set-up for observing Newton's rings is shown in the Figure. A plano-

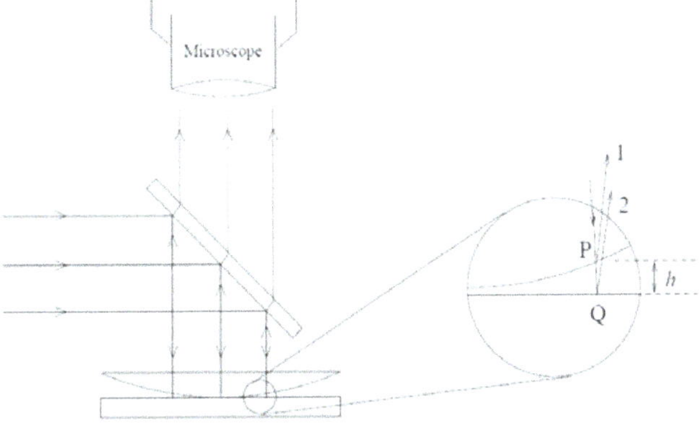

Set up for Newton's rings

convex lens is placed on a flat glass plate as shown in the figure. The radius of curvature of the plano-convex lens is large (50–100cm). This makes a very thin air film between the lower surface of the lens and the upper surface of the glass plate. A monochromatic light (like sodium light) enters from the left and is incident on a second glass plate, which is making an angle of 45° with the vertical. One could also use white light like sunlight instead of monochromatic light. For sunlight one would observe coloured fringes. This inclined glass plate, reflects the light down on the plano-convex lens. Now any ray incident on the plano-convex lens goes through multiple reflections and in this process its amplitude gets divided. The interference rings are produced by the superposition of ray 1 and ray 2 which are reflected from the lower surface of the plano-convex (at point P) and upper surface of the horizontal glass plate (at point Q) respectively as shown in Figure. There are couple of reflections more, viz. one from the upper surface of the plano-convex lens and the other from the lower surface of the horizontal glass plate, but they are not of much concern as they are not coherent due to the thickness of the lens and the glass plate. These rays produce a monotonous background of uniform illumination.

Rays 1 and 2 are coherent to each other and they travel upwards through the inclined glass plate and the interference pattern is observed through the microscope. In this system one observes circular fringes due to the circular symmetry of the lens around the point of its contact with the horizontal glass plate. Like the double slit we can calculate the path difference (or the phase difference) between the rays 1 and 2 and obtain the condition for maxima and minima. If the vertical distance between the points P and Q is taken as h then the ray two lags by a path of $2h$ amounting to a phase difference of $4\pi h/\lambda$. To this we have to add another phase difference of π (or path difference of $\lambda/2$) since the ray 2 suffers an additional difference of phase π for travelling from a rarer medium (air) and falling on a denser medium (glass) and getting reflected. Now using the set-up geometry we can find the radius of the nth

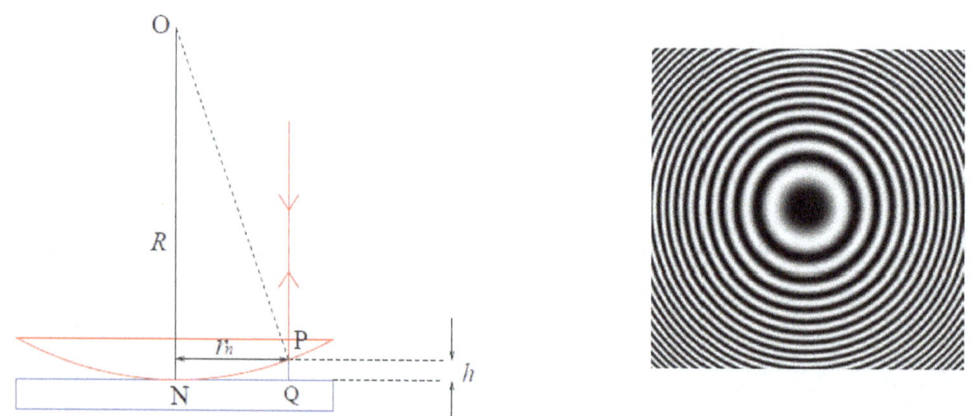

Set-up geometry and Newton's rings

bright or dark ring in terms of radius of curvature of the lens and wavelength of the incident light. Let R be the radius of curvature of the plano-convex lens and wavelength of the light falling on it be λ. From Figure we have,

$$(R-h)^2 + r_n^2 = R^2 \Rightarrow 2h = \frac{r_n^2}{R}$$

We have neglected the term h^2 in comparison to other two terms since $R >> h.$. The conditions for maxima and minima now become, $2h = \left(n - \frac{1}{2}\right)\lambda$ and $2h = n\lambda$ respectively. So

$$r_n = \sqrt{(n - \frac{1}{2})R\lambda} \text{ for a maximum; bright ring}$$

$$= \sqrt{nR\lambda} \text{ for a minimum; dark ring, n=1,2,3...}$$

By measuring various diameters of the different rings with the help of a travelling microscope one can measure the wavelength of the monochromatic light if the radius of curvature of the plno-convex lens is known. Newton's rings set up has several applications. It is a useful device for calculating refractive

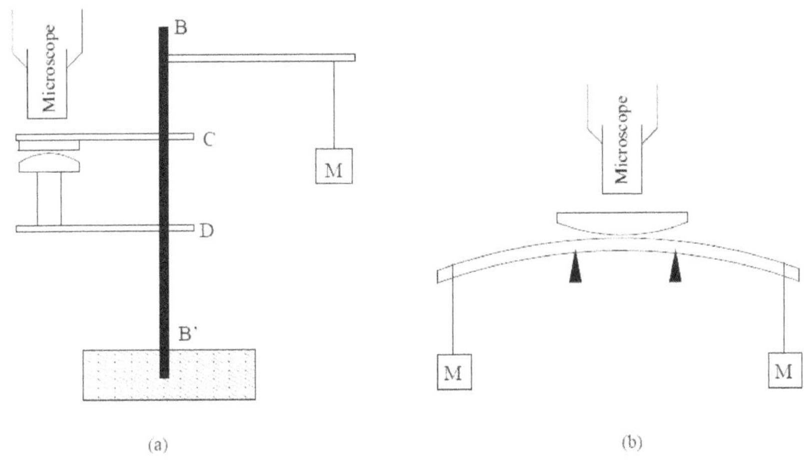

(a) Searle's apparatus for Young's modulus (b) Cornu's method

indices of various transparent liquids. In this method one introduces a few drops of the liquid between the plano-convex lens and the glass plate. This converts the air film into a thin liquid film and increases the path difference $2h$ by a factor of μ (the refractive index of the the liquid). Measuring the diameters of the rings, knowing the radius of curvature of the lens and wavelength of the monochromatic light one finds the refractive index of the liquid very accurately. Newton's rings can be useful in measuring Young's modulus of metals using Searle's apparatus. The sketch of the Searle's apparatus is shown in Figure (a). Due to the mass M the metallic rod BB' will be strained. The strain in the portion CD of the rod will cause the separation between the lens and the glass plate to change. And this change in turn will shrink the fringes and they will slowly disappear in the centre as we increase the mass slowly. The reduction of mass will cause the fringes to appear from the centre. Counting these appearance or disappearance of fringes one has a measure of strain in units of wavelength λ. Newton's rings can also be used to measure the Young's modulus and Poisson ratio for glass or other transparent materials. In this case one uses Cornu's method. In this the lower transparent plate is long and is kept on a couple of knife edges. It is strained by attaching masses at both ends as shown in the Figure, The thickness of the air film would change slowly as one increases the attached masses. In this case the circular rings will be slowly deformed to elliptic rings of varying eccentricity. Again these deviation from the circular rings can be measured accurately to estimate Young's modulus and Poisson ratio for the material of the plate.

Thin-film Interference

A colorful interference pattern is observed when light is reflected from the top and bottom boundaries of a thin oil film.

Colors in light reflected from a soap bubble

Thin-film interference is a natural phenomenon in which light waves reflected by the upper and lower boundaries of a thin film interfere with one another, either enhancing or reducing the reflected light. When the thickness of the film is an odd multiple of one quarter-wavelength of the light on it, the reflected waves from both surfaces interfere to cancel each other. Since the wave cannot be reflected, it is completely transmitted instead. When the thickness is a multiple of a half-wavelength of the light, the two reflected waves reinforce each other, increasing the reflection and reducing the transmission. Thus when white light, which consists of a range of wavelengths, is incident on the film, certain wavelengths (colors) are intensified while others are attenuated. Thin-film interference explains the multiple colors seen in light reflected from soap bubbles and oil films on water. It also is the mechanism behind the action of antireflection coatings used on glasses and camera lenses.

Studying the light reflected or transmitted by a thin film can reveal information about the thickness of the film or the effective refractive index of the film medium. Thin films have many commercial applications including anti-reflection coatings, mirrors, and optical filters.

Theory

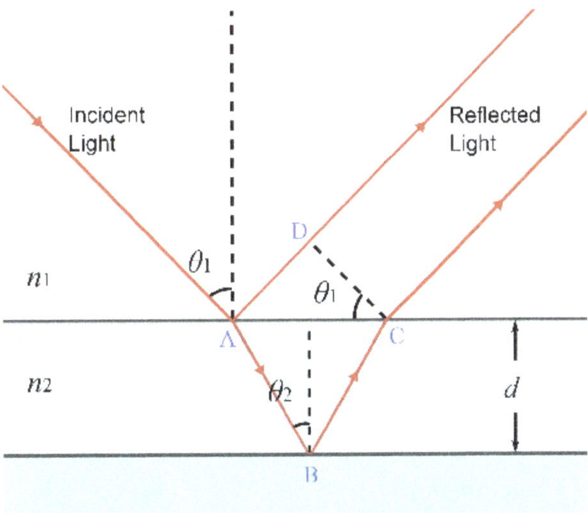

Demonstration of the optical path length difference for light reflected from the upper and lower boundaries of a thin film.

Thin-film interference caused by ITO defrosting coating on an Airbus cockpit window.

A thin film is a layer of material with thickness in the sub-nanometer to micron range. As light strikes the surface of a film it is either transmitted or reflected at the upper surface. Light that is transmitted reaches the bottom surface and may once again be transmitted or reflected. The Fresnel equations provide a quantitative description of how much of the light will be transmitted or reflected at an interface. The light reflected from the upper and lower surfaces will interfere. The degree of constructive or destructive interference between the two light waves depends on the difference in their phase. This difference in turn depends on the thickness of the film layer, the refractive index of the film, and the angle of incidence of the original wave on the film. Additionally, a phase shift of 180° or π radians may be introduced upon reflection at a boundary depending on the refractive indices of the materials on either side of the boundary. This phase shift occurs if the refractive index of the medium the light is travelling through is less than the refractive index of the material it is striking. In other words, if $n_1 < n_2$ and the light is travelling from material 1 to material 2, then a phase shift occurs upon reflection. The pattern of light that results from this interference can appear either as light and dark bands or as colorful bands depending upon the source of the incident light.

Consider light incident on a thin film and reflected by both the upper and lower boundaries. The optical path difference (OPD) of the reflected light must be calculated in order to determine the condition for interference. Referring to the ray diagram above, the OPD between the two waves is the following:

$$OPD = n_2(\overline{AB} + \overline{BC}) - n_1(\overline{AD})$$

Where,

$$\overline{AB} = \overline{BC} = \frac{d}{\cos(\theta_2)}$$

$$\overline{AD} = 2d\tan(\theta_2)\sin(\theta_1)$$

Using Snell's Law, $n_1 \sin(\theta_1) = n_2 \sin(\theta_2)$

$$OPD = n_2\left(\frac{2d}{\cos(\theta_2)}\right) - 2d\tan(\theta_2)n_2\sin(\theta_2)$$

$$OPD = 2n_2 d\left(\frac{1-\sin^2(\theta_2)}{\cos(\theta_2)}\right)$$

$$OPD = 2n_2 d\cos(\theta_2)$$

Interference will be constructive if the optical path difference is equal to an integer multiple of the wavelength of light, λ.

$$2n_2 d\cos(\theta_2) = m\lambda$$

This condition may change after considering possible phase shifts that occur upon reflection.

Monochromatic Source

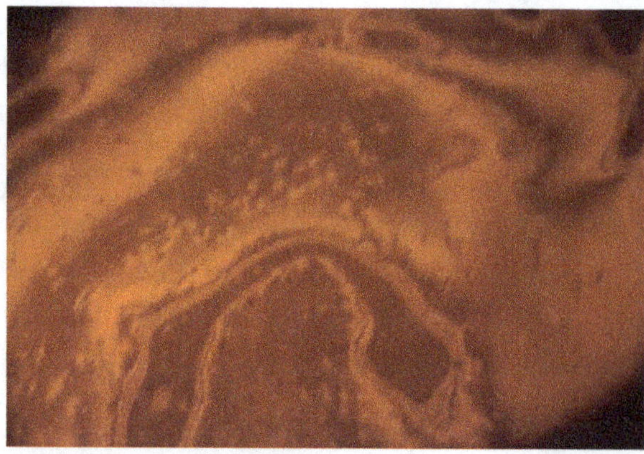

Gasoline on water shows a pattern of bright and dark fringes when illuminated with 589nm laser light.

Where incident light is monochromatic in nature, interference patterns appear as light and dark bands. Light bands correspond to regions at which constructive interference is occurring between the reflected waves and dark bands correspond to destructive interference regions. As the thickness of the film varies from one location to another, the interference may change from constructive to destructive. A good example of this phenomenon, termed "Newton's rings," demonstrates the interference pattern that results when light is reflected from a spherical surface adjacent to a flat surface. Concentric rings are observed when the surface is illuminated with monochromatic light. This phenomenon is used with optical flats to measure the shape and flatness of surfaces.

Broadband Source

If the incident light is broadband, or white, such as light from the sun, interference patterns appear as colorful bands. Different wavelengths of light create constructive interference for different film thicknesses. Different regions of the film appear in different colors depending on the local film thickness.

Phase Interaction

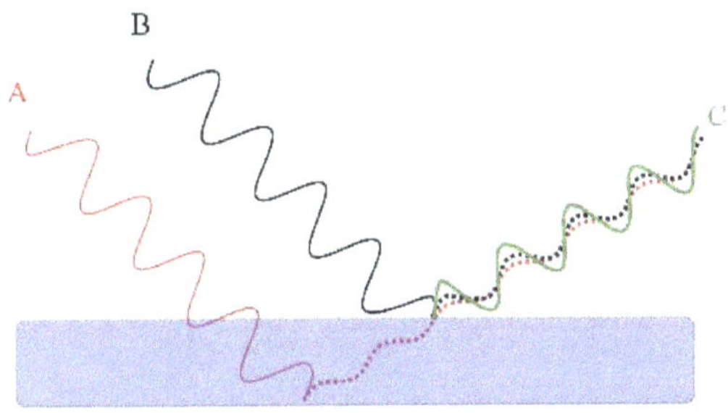

Constructive phase interaction

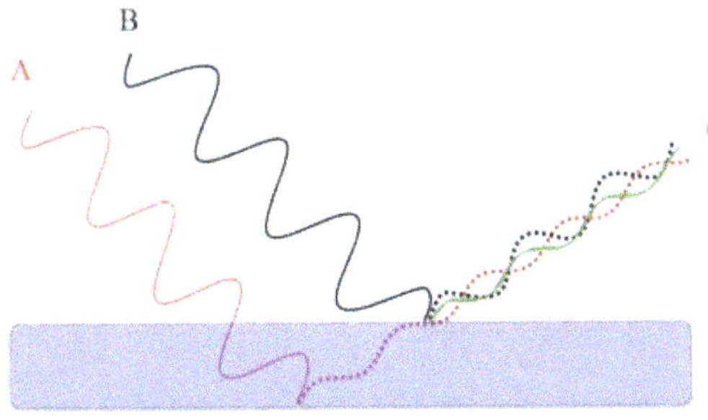

Destructive phase interaction

This section provides a simplified explanation of the phase relationship responsible for most of this phenomenon. The figures show two incident light beams (A and B). Each beam produces a reflected beam (dashed). The reflections of interest are beam A's reflection off of the lower surface and beam B's reflection off of the upper surface. These reflected beams combine to produce a resultant beam (C). If the reflected beams are in phase (as in the first figure) the resultant beam is relatively strong. If, on the other hand, the reflected beams have opposite phase, the resulting beam is attenuated (as in the second figure).

The phase relationship of the two reflected beams depends on the relationship between the wavelength of beam A in the film, and the film's thickness. If the total distance beam A travels in the film is an integer multiple of the wavelength of the beam in the film, then the two reflected beams are in phase and constructively interfere (as depicted in the first figure). If the distance traveled by beam A is an odd integer multiple of the half wavelength of light in the film, the beams destructively interfere (as in the second figure). Thus, the film shown in these figures reflects more strongly at the wavelength of the light beam in the first figure, and less strongly at that of the beam in the second figure.

Examples

The type of interference that occurs when light is reflected from a thin film is dependent upon the wavelength and angle of the incident light, the thickness of the film, the refractive indices of the material on either side of the film, and the index of the film medium. Various possible film configurations and the related equations are explained in more detail in the examples below.

Soap Bubble

Thin film interference in a soap bubble. Color varies with film thickness.

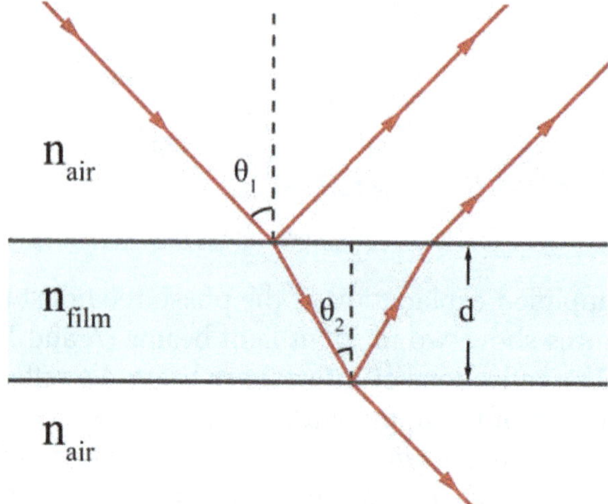

Light incident on a soap film in air

In the case of a soap bubble, light travels through air and strikes a soap film. The air has a refractive index of 1 ($n_{air} = 1$) and the film has an index that is larger than 1 ($n_{film} > 1$). The reflection that occurs at the upper boundary of the film (the air-film boundary) will introduce a 180° phase shift in the reflected wave because the refractive index of the air is less than the index of the film ($n_{air} < n_{film}$). Light that is transmitted at the upper air-film interface will continue to the lower film-air interface where it can be reflected or transmitted. The reflection that occurs at this boundary will not change the phase of the reflected wave because $n_{film} > n_{air}$. The condition for interference for a soap bubble is the following:

$$2n_{film} d\cos(\theta_2) = \left(m - \frac{1}{2}\right)\lambda$$ for constructive interference of reflected light

$$2n_{film} d\cos(\theta_2) = m\lambda$$ for destructive interference of reflected light

Where d is the film thickness, n_{film} is the refractive index of the film, θ_2 is the angle of incidence of the wave on the lower boundary, m is an integer, and λ is the wavelength of light.

Oil Film

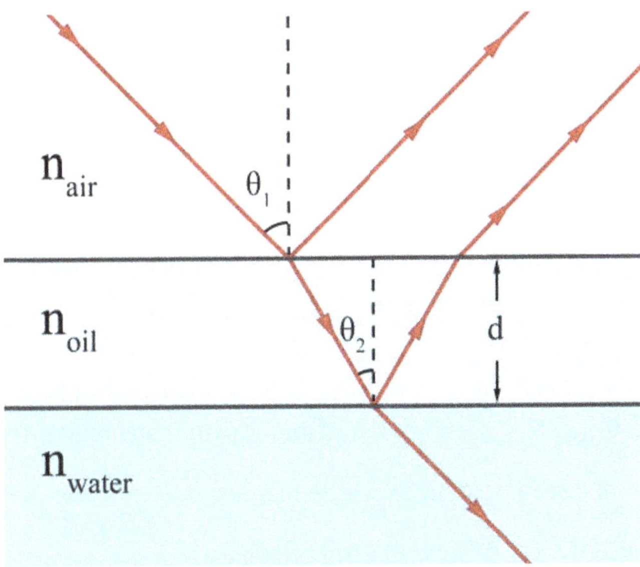

Light incident on an oil film on water

In the case of a thin oil film, a layer of oil sits on top of a layer of water. The oil may have an index of refraction near 1.5 and the water has an index of 1.33. As in the case of the soap bubble, the materials on either side of the oil film (air and water) both have refractive indices that are less than the index of the film. $n_{air} < n_{water} < n_{oil}$. There will be a phase shift upon reflection from the upper boundary because $n_{air} < n_{oil}$ but no shift upon reflection from the lower boundary because $n_{oil} > n_{water}$. The equations for interference will be the same.

$$2n_{oil} d\cos(\theta_2) = \left(m - \frac{1}{2}\right)\lambda$$ for constructive interference of reflected light

$$2n_{oil} d\cos(\theta_2) = m\lambda$$ for destructive interference of reflected light

Anti-reflection Coatings

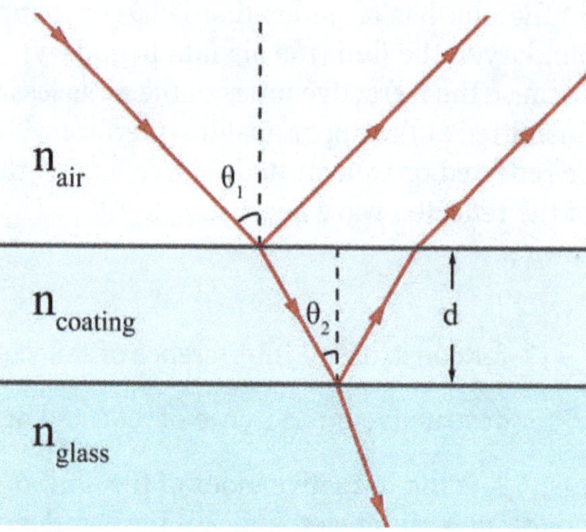

Light incident on an anti-reflection coating on glass

An anti-reflection coating eliminates reflected light and maximizes transmitted light in an optical system. A film is designed such that reflected light produces destructive interference and transmitted light produces constructive interference for a given wavelength of light. In the simplest implementation of such a coating, the film is created so that its optical thickness dn_{coating} is a quarter-wavelength of the incident light and its refractive index is greater than the index of air and less than the index of glass.

$$n_{\text{air}} < n_{\text{coating}} < n_{\text{glass}}$$

$$d = \lambda / (4n_{\text{coating}})$$

A 180° phase shift will be induced upon reflection at both the top and bottom interfaces of the film because $n_{\text{air}} < n_{\text{coating}}$ and $n_{\text{coating}} < n_{\text{glass}}$. The equations for interference of the reflected light are:

$2n_{\text{coating}} d\cos(\theta_2) = m\lambda$ for constructive interference

$2n_{\text{coating}} d\cos(\theta_2) = \left(m - \frac{1}{2}\right)\lambda$ for destructive interference

If the optical thickness dn_{coating} is equal to a quarter-wavelength of the incident light and if the light strikes the film at normal incidence $(\theta_2 = 0)$, the reflected waves will be completely out of phase and will destructively interfere. Further reduction in reflection is possible by adding more layers, each designed to match a specific wavelength of light.

Interference of transmitted light is completely constructive for these films.

In Nature

Structural coloration due to thin-film layers is common in the natural world. The wings of many insects act as thin films because of their minimal thickness. This is clearly visible in the wings of many flies and wasps. In butterflies, the thin-film optics are visible when the wing itself is not covered by pigmented wing scales, which is the case in the blue wing spots of the *Aglais io* butterfly.

The glossy appearance of buttercup flowers is also due to a thin film as well as the shiny breast feathers of the bird of paradise.

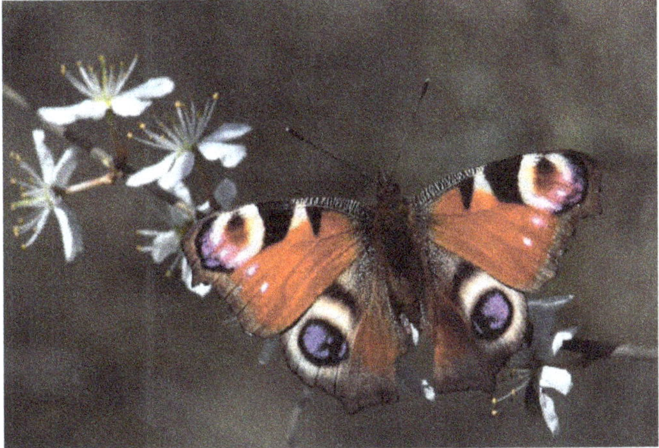

The blue wing patches of the *Aglais io* butterfly are due to thin-film interference.

Applications

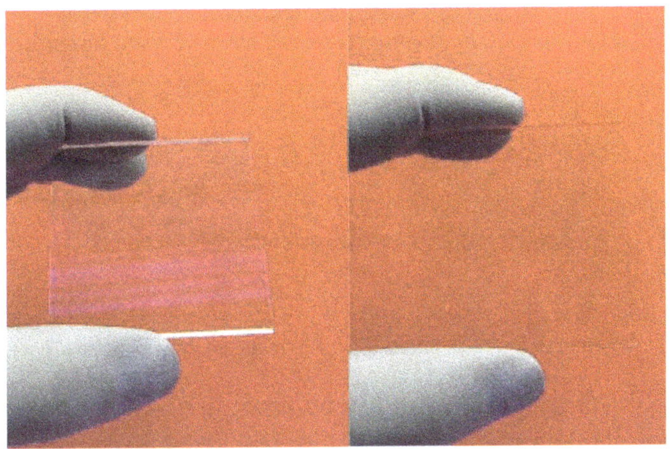

An antireflection-coated optical window. At a 45° angle the coating is slightly thicker to the incident light, causing the center wavelength to shift toward the red and reflections appear at the purple end of the spectrum. At 0°, for which this coating was designed, almost no reflection is observed.

Thin films are used commercially in anti-reflection coatings, mirrors, and optical filters. They can be engineered to control the amount of light reflected or transmitted at a surface for a given wavelength. A Fabry–Pérot etalon takes advantage of thin film interference to selectively choose which wavelengths of light are allowed to transmit through the device. These films are created through deposition processes in which material is added to a substrate in a controlled manner. Methods include chemical vapor deposition and various physical vapor deposition techniques.

Thin films are also found in nature. Many animals have a layer of tissue behind the retina, the Tapetum lucidum, that aids in light collecting. The effects of thin-film interference can also be seen in oil slicks and soap bubbles. The reflectance spectrum of a thin-film features distinct oscillations and the extrema of the spectrum can be used to calculate the thickness of the thin-film.

Ellipsometry is a technique that is often used to measure properties of thin films. In a typical ellipsometry experiment polarized light is reflected off a film surface and is measured by a detector. The complex reflectance ratio, ρ, of the system is measured. A model analysis in then conducted in which this information is used to determine film layer thicknesses and refractive indices.

Dual polarisation interferometry is an emerging technique for measuring refractive index and thickness of molecular scale thin films and how these change when stimulated.

History

The gloss of buttercup flowers is due to thin-film interference.

Tempering colors are produced when steel is heated and a thin film of iron oxide forms on the surface. The color indicates the temperature the steel reached, which made this one of the earliest practical uses of thin-film interference.

Iridescence caused by thin-film interference is a commonly observed phenomenon in nature, being found in a variety of plants and animals. One of the first known studies of this phenomenon was conducted by Robert Hooke in 1665. In *Micrographia*, Hooke postulated that the iridescence in peacock feathers was caused by thin, alternating layers of plate and air. In 1704, Isaac Newton stated in his book, *Opticks*, that the iridescence in a peacock feather was due to the fact that the transparent layers in the feather were so thin. In 1801, Thomas Young provided the first explanation of constructive and destructive interference. Young's contribution went largely unnoticed until the work of Augustin Fresnel. In 1816, Fresnel helped to establish the wave theory of light. However, very little explanation could be made of the iridescence until the 1870s, when James Maxwell and Heinrich Hertz helped to explain the electromagnetic nature of light. After the invention of the Fabry–Perot interferometer, in 1899, the mechanisms of thin-film interference could be demonstrated on a larger scale.

In much of the early work, scientists tried to explain iridescence, in animals like peacocks and scarab beetles, as some form of surface color, such as a dye or pigment that might alter the light when reflected from different angles. In 1919, Lord Rayleigh proposed that the bright, changing colors were not caused by dyes or pigments, but by microscopic structures, which he termed "structural colors." In 1923, C. W. Mason noted that the barbules in the peacock feather were made from very thin layers. Some of these layers were colored while others were transparent. He noticed that pressing the barbule would shift the color toward the blue, while swelling it with a chemical would shift it toward the red. He also found that bleaching the pigments from the feathers did not remove the iridescence. This helped to dispel the surface color theory and reinforce the structural color theory.

In 1925, Ernest Merritt, in his paper *A Spectrophotometric Study of Certain Cases of Structural Color*, first described the process of thin-film interference as an explanation for the iridescence. The first examination of iridescent feathers by an electron microscope occurred in 1939, revealing complex thin-film structures, while an examination of the morpho butterfly, in 1942, revealed an extremely tiny array of thin-film structures on the nanometer scale.

The first production of thin-film coatings occurred quite by accident. In 1817, Joseph Fraunhofer discovered that, by tarnishing glass with nitric acid, he could reduce the reflections on the surface. In 1819, after watching a layer of alcohol evaporate from a sheet of glass, Fraunhofer noted that colors appeared just before the liquid evaporated completely, deducing that any thin film of transparent material will produce colors.

Little advancement was made in thin-film coating technology until 1936, when John Strong began evaporating fluorite in order to make anti-reflection coatings on glass. During the 1930s, improvements in vacuum pumps made vacuum deposition methods, like sputtering, possible. In 1939, Walter H. Geffcken created the first interference filters using dielectric coatings.

Multiple Beam interference: Thin Films

In this section we will study thin films. The applications of thin films in various industries are countless. Here we will restrict ourselves in optical properties of thin films. The simplest application of a metallic thin film is found in a household mirror which makes a the glass almost perfectly reflecting. On rainy days often when the road is wet we observe beautiful coloured patterns on various areas patches of the road. The coloured patterns are seen due to the interference in thin films. Actually when a few drops of diesel or petrol from the vehicles fall on wet roads they spread as thin oil films (a few microns thick). Since the oil is lighter than the water it remains on the top. The white light from the sky or the Sun reflects through this film giving vibrant colours. Usually the thickness of the oil films formed in such a way is non-uniform. Different interference conditions (a maximum, intermediate or a minimum) are satisfied for different thicknesses for the same wavelength like the case of Newton's rings with air film. Even for the same thickness different conditions are satisfied for different wavelengths since in white light one has continuous wavelengths. A single colour is observed as a patch where the thickness of the film is uniform.

Stokes Relations

Consider a light ray incident on a glass surface obliquely with an amplitude a. Let us take the reflection coefficient be r and transmission coefficient be t for air to glass. Further, assume the reflection and transmission coefficients as r' and t' respectively for glass to air. This is depicted in Figure. Now the initial amplitude, a, splits as ar for the reflected ray and at for the refracted ray (shown in blue). Next we use the reversibility of the optical path to construct back the original ray by reversing the paths of reflected and refracted rays. When we reverse the path of reflected ray, ar, we have one reflected ray ar^2 and one refracted ray art (these are shown in red). Next reversing the refracted

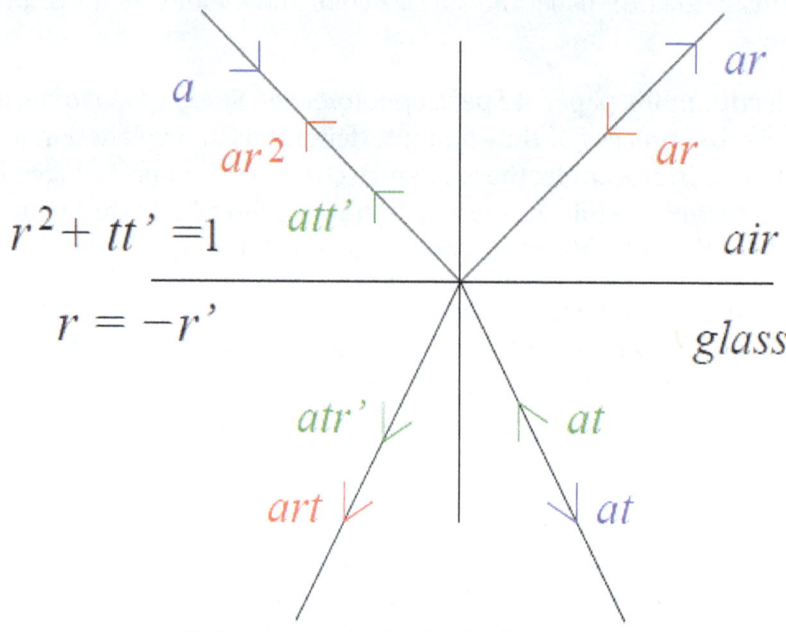

Stokes parameters for air-glass interface

ray, at, we have two more rays, viz. one reflected atr' and one refracted att' (shown in green). Now if we add all these amplitudes we must have a null. So, we have the following conditions,

$$a = a(r^2 + tt') \Rightarrow r^2 + tt' = 1 \text{ and } art + atr' = 0 \Rightarrow r = -r',$$

on reflection and refraction coefficients. These relations are known as Stokes relations. The second of the above relations, (16.1), shows that there is a relative phase difference of π $\left(\text{since } e^{i\pi} = -1\right)$, between the two coefficients r and r'. This is the extra phase we considered in the earlier section for a reflection in a rarer medium from a interface backed by a denser medium.

Experiments show that indeed this the case.

Uniform Thin Films

We will now obtain the conditions for maxima and minima for a uniform thin film of thickness d and refractive index μ. We consider the situation shown in Figure. The figure is self explanatory. A ray is falling obliquely (with angle incidence of i) on a uniform thin film of refractive index μ, There are multiple reflections inside the film and we have many reflected and transmitted rays. Now if these rays are focussed with the help of a converging lens one would observe a blob of dark or bright light depending one whether these rays interfere destructively or constructively. We first find the path difference, Δ, between the ray 2 and ray 1. We make the following constructions in Figure before proceeding. RM is a perpendicular on ray 1. ON is a perpendicular on RS. SP is a straight and equal extension of RS.

$$\Delta = \left[(OS+SR)(in\ film)\right] - \left[OM(in\ air)\right]$$
$$= \left[(PS+SR)(in\ film)\right] - \left[OM(in\ air)\right]$$
$$= \left[PR(in\ film)\right] - \left[OM(in\ air)\right]$$

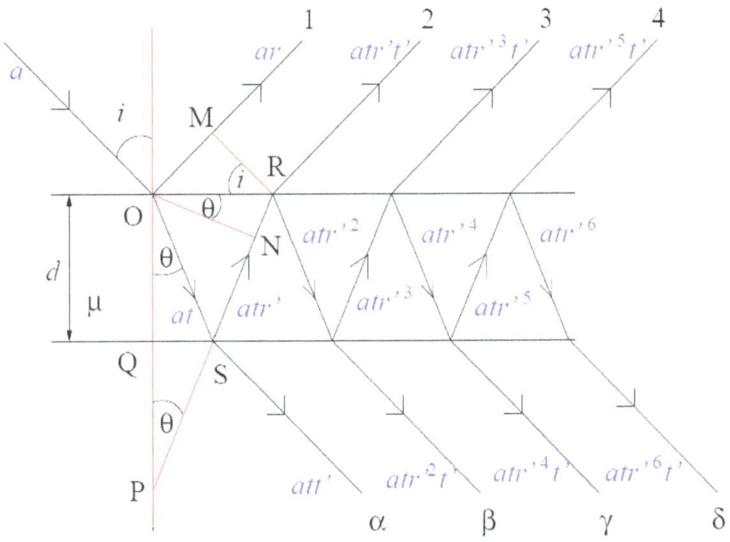

A ray incident on a thin film obliquely

$$= \mu(PN+NR) - OM = \mu(PN) + \sin i \frac{1}{\sin\theta} NR - OM$$
$$= \mu(PN) + \frac{OM}{OR}\frac{OR}{NR} NR - OM = \mu(PN) + OM - OM$$
$$= \mu(PN) = \mu(OP\cos\theta)$$
$$= 2\mu d\cos\theta$$

Case-I: $\Delta = m\lambda$

Now if $\Delta = m\lambda$, m, 0, 1, 2, · · ·, the rays 2, 3, 4, · · · are in phase with one another.

So one can easily sum the the reflected rays 2,3,4 etc. leaving the ray 1,

$$2+3+4+\cdots = atr't' + atr'^3t' + atr'^5t'\cdots = \frac{atr't'}{1-r'^2} = ar',$$

where we have used the second Stokes relation. The magnitude of this sum is equal to the amplitude of ray 1. But the ray 1 is out of phase with this sum since it suffers an extra reflection at O from rarer to denser medium causing a π phase difference. The situation is equivalent to superposition of two rays having equal amplitude and opposite phase. So we get a minimum or a dark blob of zero intensity for the reflected rays in this situation, viz. $\Delta = m\lambda$. In this situation all the transmitted rays, $\alpha,\ \beta,\ \gamma,\ \delta,\cdots$ are in phase. Like the reflected rays we can add them up,

$$\alpha+\beta+\gamma+\delta+\cdots=att'\left(1+r'^2+r'^4+r'^6+\cdots\right)=att'/\left(1-r'^2\right)=a.$$

This gives a maximum of original intensity for the transmitted rays. This can be appreciated intuitively by energy conservation, since there is no reflected ray so the whole incident amplitude is transmitted.

Case-II: $\Delta=(m+\ 1/2)\lambda$

When the path difference $\Delta=\ (m+\ 1/2)\lambda$, the rays 1, 2, 4, 6, . . . are in phase with one another, whereas the rays 3, 5, 7, . . . are in phase among themselves but out of phase with 1, 2, 4, 6,.... Similarly the rays $\alpha,\ \gamma$, ... are added up in phase and $\beta,\ \delta,\ldots$ are also added up in phase separately, but these two bunches are out of phase with each other. This situation produces a maximum for the reflected rays since the two interfering bunches have different amplitudes and the first bunch, i.e. $1+2+4+\cdots$ is always stronger than $3\ +\ 5\ +\ 7\ +\cdots$, since each member of the fist sum is individually stronger than corresponding member of the second sum. The same condition produces a minimum (but non zero intensity) for the transmitted rays due to the cancellation of two out of phase bunches $\alpha,\ \gamma$,... and $\beta,\ \delta$,....

Now we are in a position to understand the colours of the oil films. For a particular thickness the film will not reflect a particular wavelength (or colour) if the appropriate conditions for destructive interference are met, and from the reflecting light the same colour would be missing and one would observe a complementary colour which is mixture of different colours with different amplitudes. So a patch of one colour in a film is created by the same thickness of the film available below the patch. A different thickens would chop off a different wavelength and there another colour would be visible (which is complementary to the missing coulor). Thin films are used as filters where one would like to reduce or chop off unwanted wavelengths of light. As we know the ultraviolet rays are harmful to eyes, the good companies use thin film coatings on their spectacles. The thickness of the coating is so adjusted that unwanted wavelengths would satisfy the maximum condition for the reflected rays and they would not enter the eyes. This can be calculated by knowing the refractive index of the coating and the wavelength of light to be eliminated.

Coherence (Physics)

In physics, two wave sources are perfectly coherent if they have a constant phase difference and the same frequency. Coherence is an ideal property of waves that enables stationary (i.e. temporally and spatially constant) interference. It contains several distinct concepts, which are limiting cases that never quite occur in reality but allow an understanding of the physics of waves, and has become a very important concept in quantum physics. More generally, coherence describes all properties of the correlation between physical quantities of a single wave, or between several waves or wave packets.

Interference is nothing more than the addition, in the mathematical sense, of wave functions. A single wave can interfere with itself, but this is still an addition of two waves. Constructive or destructive interferences are limit cases, and two waves always interfere, even if the result of the addition is complicated or not remarkable.

When interfering, two waves can add together to create a wave of greater amplitude than either one (constructive interference) or subtract from each other to create a wave of lesser amplitude than either one (destructive interference), depending on their relative phase. Two waves are said to be coherent if they have a constant relative phase. The amount of coherence can readily be measured by the interference visibility, which looks at the size of the interference fringes relative to the input waves (as the phase offset is varied); a precise mathematical definition of the degree of coherence is given by means of correlation functions.

Spatial coherence describes the correlation (or predictable relationship) between waves at different points in space, either lateral or longitudinal. Temporal coherence describes the correlation between waves observed at different moments in time. Both are observed in the Michelson–Morley experiment and Young's interference experiment. Once the fringes are obtained in the Michelson interferometer, when one of the mirrors is moved away gradually, the time for the beam to travel increases and the fringes become dull and finally are lost, showing temporal coherence. Similarly, if in a double-slit experiment, the space between the two slits is increased, the coherence dies gradually and finally the fringes disappear, showing spatial coherence. In both cases, the fringe amplitude slowly disappears, as the path difference increases past the coherence length.

Introduction

Coherence was originally conceived in connection with Thomas Young's double-slit experiment in optics but is now used in any field that involves waves, such as acoustics, electrical engineering, neuroscience, and quantum mechanics. The property of coherence is the basis for commercial applications such as holography, the Sagnac gyroscope, radio antenna arrays, optical coherence tomography and telescope interferometers (astronomical optical interferometers and radio telescopes).

Mathematical Definition

A precise definition is given at degree of coherence.

The coherence function between two signals $x(t)$ and $y(t)$ is defined as

$$\gamma_{xy}^2(f) = \frac{|S_{xy}(f)|^2}{S_{xx}(f)S_{yy}(f)}$$

where $S_{xy}(f)$ is the cross-spectral density of the signal and $S_{xx}(f)$ and $S_{yy}(f)$ are the power spectral density functions of $x(t)$ and $y(t)$, respectively. The cross-spectral density and the power spectral density are defined as the Fourier transforms of the cross-correlation and the autocorrelation signals, respectively. For instance, if the signals are functions of time, the cross-correlation is a measure of the similarity of the two signals as a function of the time lag relative to each other and the autocorrelation is a measure of the similarity of each signal with itself in different instants of time. In this case the coherence is a function of frequency. Analogously, if $x(t)$ and $y(t)$ are functions of space, the cross-correlation measures the similarity of two signals in different points in space and the autocorrelations the similarity of the signal relative to itself for a certain separation distance. In that case, coherence is a function of wavenumber (spatial frequency).

The coherence varies in the interval $0 \leqslant \gamma_{xy}^2(f) \leqslant 1$.. If $\gamma_{xy}^2(f)=1$ it means that the signals are perfectly correlated or linearly related and if $\gamma_{xy}^2(f)=0$ they are totally uncorrelated. If a linear system is being measured, $x(t)$ being the input and $y(t)$ the output, the coherence function will be unitary all over the spectrum. However, if non-linearities are present in the system the coherence will vary in the limit given above.

Coherence and Correlation

The coherence of two waves expresses how well correlated the waves are as quantified by the cross-correlation function. The cross-correlation quantifies the ability to predict the phase of the second wave by knowing the phase of the first. As an example, consider two waves perfectly correlated for all times. At any time, phase difference will be constant. If, when combined, they exhibit perfect constructive interference, perfect destructive interference, or something in-between but with constant phase difference, then it follows that they are perfectly coherent. As will be discussed below, the second wave need not be a separate entity. It could be the first wave at a different time or position. In this case, the measure of correlation is the autocorrelation function (sometimes called self-coherence). Degree of correlation involves correlation functions.

Examples of Wave-like States

These states are unified by the fact that their behavior is described by a wave equation or some generalization thereof.

- Waves in a rope (up and down) or slinky (compression and expansion)
- Surface waves in a liquid
- Electromagnetic signals (fields) in transmission lines
- Sound
- Radio waves and Microwaves
- Light waves (optics)
- Electrons, atoms and any other object (such as a baseball), as described by quantum physics

In most of these systems, one can measure the wave directly. Consequently, its correlation with another wave can simply be calculated. However, in optics one cannot measure the electric field directly as it oscillates much faster than any detector's time resolution. Instead, we measure the intensity of the light. Most of the concepts involving coherence which will be introduced below were developed in the field of optics and then used in other fields. Therefore, many of the standard measurements of coherence are indirect measurements, even in fields where the wave can be measured directly.

Temporal Coherence

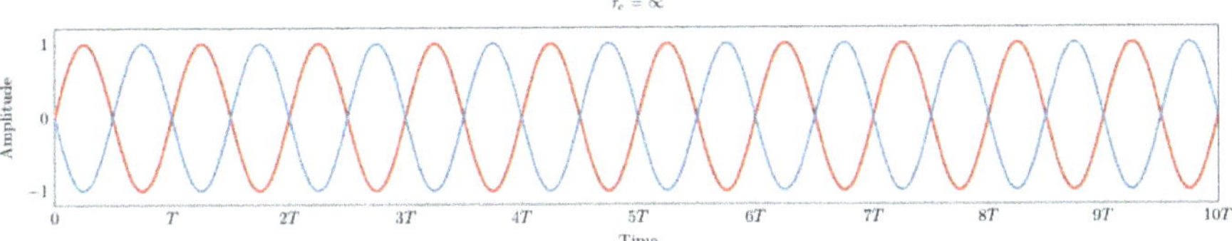

The amplitude of a single frequency wave as a function of time *t* (red) and a copy of the same wave delayed by τ (blue). The coherence time of the wave is infinite since it is perfectly correlated with itself for all delays τ.

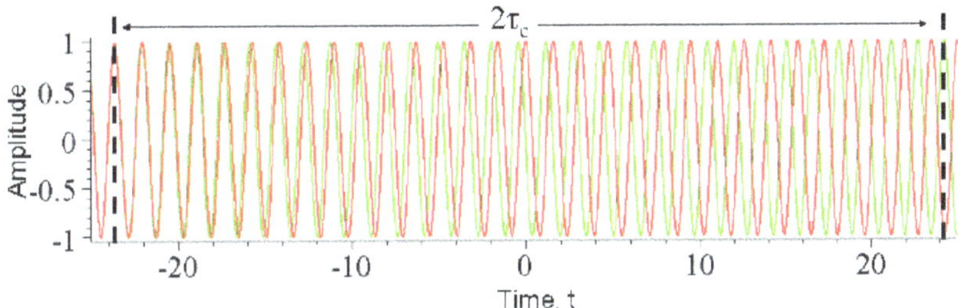

The amplitude of a wave whose phase drifts significantly in time τ_c as a function of time *t* (red) and a copy of the same wave delayed by $2\tau_c$(green). At any particular time t the wave can interfere perfectly with its delayed copy. But, since half the time the red and green waves are in phase and half the time out of phase, when averaged over t any interference disappears at this delay.

Temporal coherence is the measure of the average correlation between the value of a wave and itself delayed by τ, at any pair of times. Temporal coherence tells us how monochromatic a source is. In other words, it characterizes how well a wave can interfere with itself at a different time. The delay over which the phase or amplitude wanders by a significant amount (and hence the correlation decreases by significant amount) is defined as the coherence time τ_c. At a delay of τ=0 the degree of coherence is perfect, whereas it drops significantly as the delay passes $\tau=\tau_c$. The coherence length L_c is defined as the distance the wave travels in time τ_c.

One should be careful not to confuse the coherence time with the time duration of the signal, nor the coherence length with the coherence area.

The Relationship between Coherence Time and Bandwidth

It can be shown that the larger the range of frequencies Δf a wave contains, the faster the wave decorrelates (and hence the smaller τ_c is). Thus there is a tradeoff:

$$\tau_c \Delta f \lesssim 1.$$

Formally, this follows from the convolution theorem in mathematics, which relates the Fourier transform of the power spectrum (the intensity of each frequency) to its autocorrelation.

Examples of Temporal Coherence

We consider four examples of temporal coherence.

- A wave containing only a single frequency (monochromatic) is perfectly correlated with itself at all time delays, in accordance with the above relation.

- Conversely, a wave whose phase drifts quickly will have a short coherence time.
- Similarly, pulses (wave packets) of waves, which naturally have a broad range of frequencies, also have a short coherence time since the amplitude of the wave changes quickly.
- Finally, white light, which has a very broad range of frequencies, is a wave which varies quickly in both amplitude and phase. Since it consequently has a very short coherence time (just 10 periods or so), it is often called incoherent.

Monochromatic sources are usually lasers; such high monochromaticity implies long coherence lengths (up to hundreds of meters). For example, a stabilized and monomode helium–neon laser can easily produce light with coherence lengths of 300 m. Not all lasers are monochromatic, however (e.g. for a mode-locked Ti-sapphire laser, $\Delta\lambda \approx$ 2 nm - 70 nm). LEDs are characterized by $\Delta\lambda \approx$ 50 nm, and tungsten filament lights exhibit $\Delta\lambda \approx$ 600 nm, so these sources have shorter coherence times than the most monochromatic lasers.

Holography requires light with a long coherence time. In contrast, optical coherence tomography uses light with a short coherence time.

Measurement of Temporal Coherence

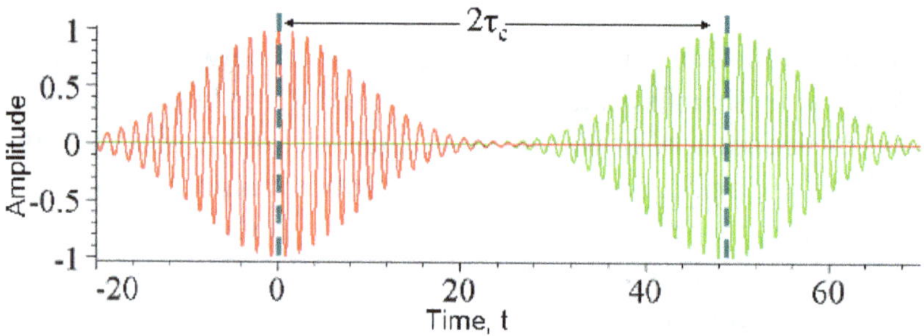

The amplitude of a wavepacket whose amplitude changes significantly in time τ_c (red) and a copy of the same wave delayed by $2\tau_c$(green) plotted as a function of time *t*. At any particular time the red and green waves are uncorrelated; one oscillates while the other is constant and so there will be no interference at this delay. Another way of looking at this is the wavepackets are not overlapped in time and so at any particular time there is only one nonzero field so no interference can occur.

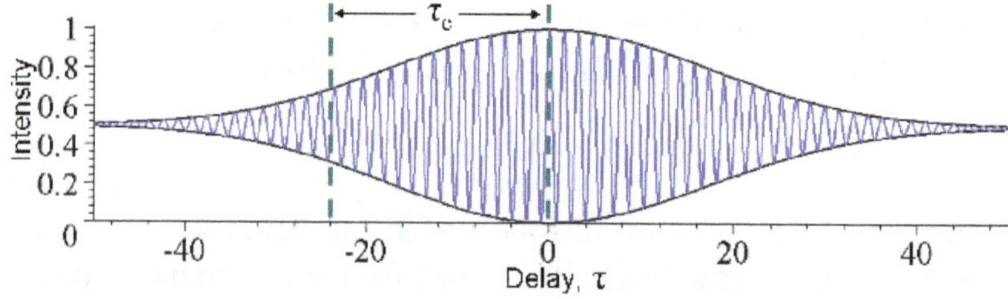

The time-averaged intensity (blue) detected at the output of an interferometer plotted as a function of delay τ for the example waves in Figures. As the delay is changed by half a period, the interference switches between constructive and destructive. The black lines indicate the interference envelope, which gives the degree of coherence. Although the waves in Figures have different time durations, they have the same coherence time.

In optics, temporal coherence is measured in an interferometer such as the Michelson interferometer or Mach–Zehnder interferometer. In these devices, a wave is combined with a copy of itself that is delayed by time τ. A detector measures the time-averaged intensity of the light exiting the interferometer. The resulting interference visibility gives the temporal coherence at delay τ. Since for most natural light sources, the coherence time is much shorter than the time resolution of any detector, the detector itself does the time averaging. Consider the example shown in Figure. At a fixed delay, here $2\tau_c$, an infinitely fast detector would measure an intensity that fluctuates significantly over a time t equal to τ_c. In this case, to find the temporal coherence at $2\tau_c$, one would manually time-average the intensity.

Spatial Coherence

In some systems, such as water waves or optics, wave-like states can extend over one or two dimensions. Spatial coherence describes the ability for two points in space, x_1 and x_2, in the extent of a wave to interfere, when averaged over time. More precisely, the spatial coherence is the cross-correlation between two points in a wave for all times. If a wave has only 1 value of amplitude over an infinite length, it is perfectly spatially coherent. The range of separation between the two points over which there is significant interference defines the diameter of the coherence area, A_c (Coherence length, often a feature of a source, is usually an industrial term related to the coherence time of the source, not the coherence area in the medium.) A_c is the relevant type of coherence for the Young's double-slit interferometer. It is also used in optical imaging systems and particularly in various types of astronomy telescopes. Sometimes people also use "spatial coherence" to refer to the visibility when a wave-like state is combined with a spatially shifted copy of itself.

Examples of Spatial Coherence

- Spatial coherence

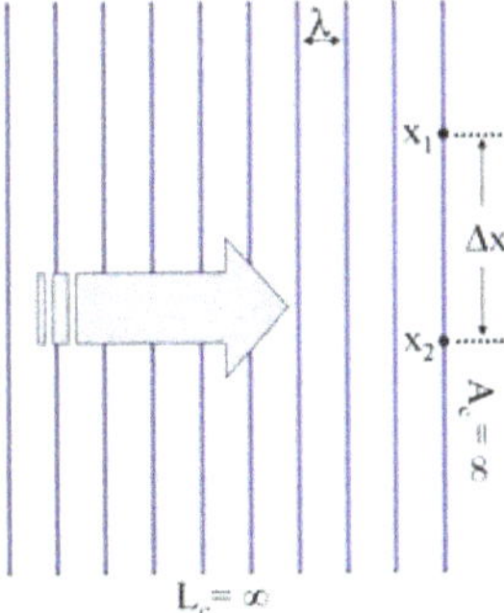

A plane wave with an infinite coherence length.

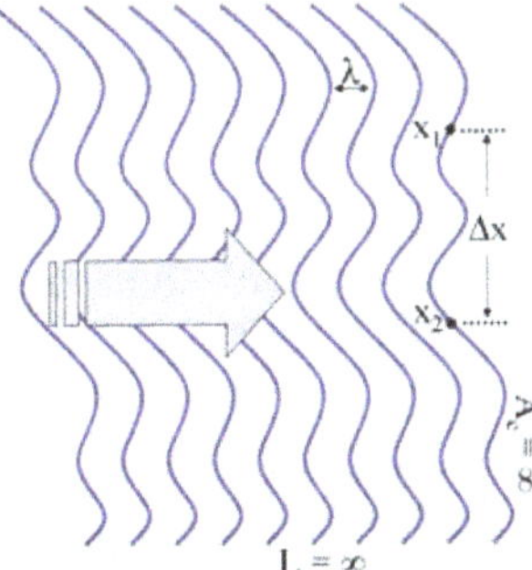

A wave with a varying profile (wavefront) and infinite coherence length.

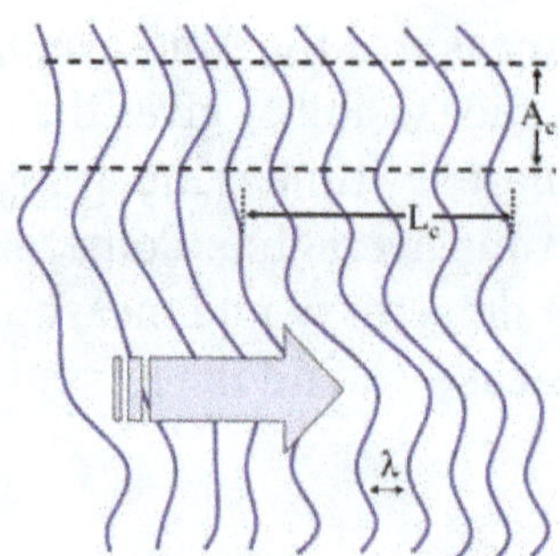

A wave with a varying profile (wavefront) and finite coherence length.

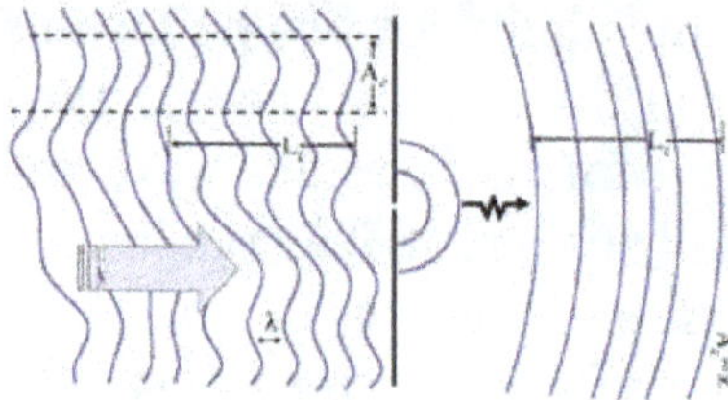

A wave with finite coherence area is incident on a pinhole (small aperture). The wave will diffract out of the pinhole. Far from the pinhole the emerging spherical wavefronts are approximately flat. The coherence area is now infinite while the coherence length is unchanged.

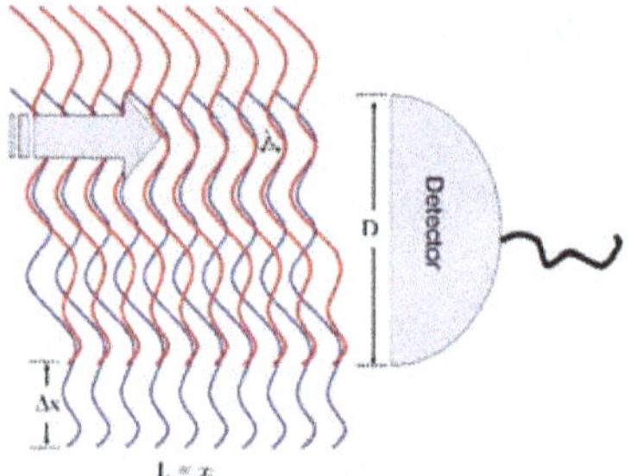

A wave with infinite coherence area is combined with a spatially shifted copy of itself. Some sections in the wave interfere constructively and some will interfere destructively. Averaging over these sections, a detector with length D will measure reduced interference visibility. For example, a misaligned Mach–Zehnder interferometer will do this.

Consider a tungsten light-bulb filament. Different points in the filament emit light independently and have no fixed phase-relationship. In detail, at any point in time the profile of the emitted light is going to be distorted. The profile will change randomly over the coherence time τ_c. Since for a white-light source such as a light-bulb τ_c is small, the filament is considered a spatially incoherent source. In contrast, a radio antenna array, has large spatial coherence because antennas at opposite ends of the array emit with a fixed phase-relationship. Light waves produced by a laser often have high temporal and spatial coherence (though the degree of coherence depends strongly on the exact properties of the laser). Spatial coherence of laser beams also manifests itself as speckle patterns and diffraction fringes seen at the edges of shadow.

Holography requires temporally and spatially coherent light. Its inventor, Dennis Gabor, produced successful holograms more than ten years before lasers were invented. To produce coherent light he passed the monochromatic light from an emission line of a mercury-vapor lamp through a pinhole spatial filter.

In February 2011 it was reported that helium atoms, cooled to near absolute zero / Bose–Einstein condensate state, can be made to flow and behave as a coherent beam as occurs in a laser.

Spectral Coherence

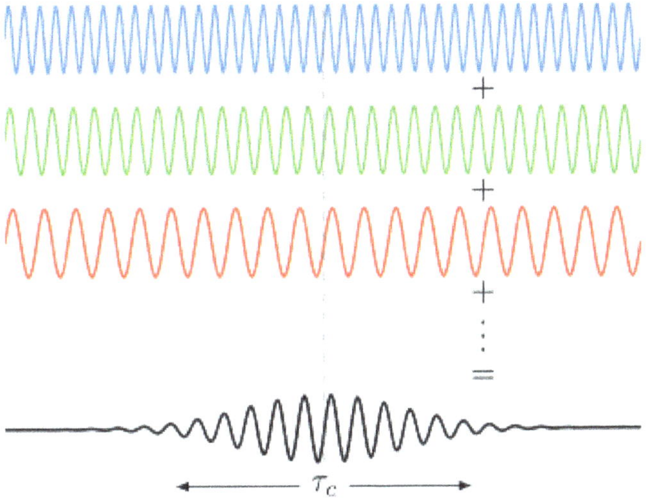

Waves of different frequencies interfere to form a localized pulse if they are coherent.

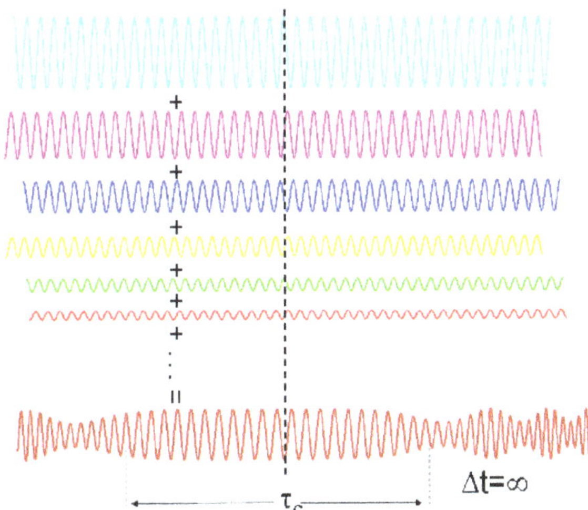

Spectrally incoherent light interferes to form continuous light with a randomly v arying phase and amplitude

Waves of different frequencies (in light these are different colours) can interfere to form a pulse if they have a fixed relative phase-relationship. Conversely, if waves of different frequencies are not coherent, then, when combined, they create a wave that is continuous in time (e.g. white light or white noise). The temporal duration of the pulse Δt is limited by the spectral bandwidth of the light Δf according to:

$$\Delta f \Delta t \geq 1,$$

which follows from the properties of the Fourier transform and results in Küpfmüller's uncertainty principle (for quantum particles it also results in the Heisenberg uncertainty principle).

If the phase depends linearly on the frequency (i.e. $\theta(f) \propto f$) then the pulse will have the minimum time duration for its bandwidth (a *transform-limited* pulse), otherwise it is chirped.

Measurement of Spectral Coherence

Measurement of the spectral coherence of light requires a nonlinear optical interferometer, such as an intensity optical correlator, frequency-resolved optical gating (FROG), or spectral phase interferometry for direct electric-field reconstruction (SPIDER).

Polarization and Coherence

Light also has a polarization, which is the direction in which the electric field oscillates. Unpolarized light is composed of incoherent light waves with random polarization angles. The electric field of the unpolarized light wanders in every direction and changes in phase over the coherence time of the two light waves. An absorbing polarizer rotated to any angle will always transmit half the incident intensity when averaged over time.

If the electric field wanders by a smaller amount the light will be partially polarized so that at some angle, the polarizer will transmit more than half the intensity. If a wave is combined with an orthogonally polarized copy of itself delayed by less than the coherence time, partially polarized light is created.

The polarization of a light beam is represented by a vector in the Poincaré sphere. For polarized light the end of the vector lies on the surface of the sphere, whereas the vector has zero length for unpolarized light. The vector for partially polarized light lies within the sphere

Applications

Holography

Coherent superpositions of *optical wave fields* include holography. Holographic objects are used frequently in daily life in bank notes and cr cards.

Non-optical Wave Fields

Further applications concern the coherent superposition of *non-optical wave fields*. In quantum mechanics for example one considers a probability field, which is related to the wave function (interpretation: density of the probability amplitude). Here the applications concern, among others, the future technologies of quantum computing and the already available technology of quantum cryptography. Additionally the problems of the following subchapter are treated.

Quantum Coherence

In quantum mechanics, all objects have wave-like properties. For instance, in Young's double-slit experiment electrons can be used in the place of light waves. Each electron's wave-function goes through both slits, and hence has two separate split-beams that contribute to the intensity pattern on a screen. According to standard wave theory these two contributions give rise to an intensity pattern of bright bands due to constructive interference, interlaced with dark bands due to destructive interference, on a downstream screen. This ability to interfere and diffract is related to coherence (classical or quantum) of the waves produced at both slits. The association of an electron with a wave is unique to quantum theory.

When the incident beam is represented by a quantum pure state, the split beams downstream of the two slits are represented as a superposition of the pure states representing each split beam. The quantum description of imperfectly coherent paths is called a mixed state. A perfectly coherent state has a density matrix (also called the "statistical operator") that is a projection onto the pure coherent state and is equivalent to a wave function, while a mixed state is described by a classical probability distribution for the pure states that make up the mixture.

Macroscopic scale quantum coherence leads to novel phenomena, the so-called macroscopic quantum phenomena. For instance, the laser, superconductivity and superfluidity are examples of highly coherent quantum systems whose effects are evident at the macroscopic scale. The macroscopic quantum coherence (Off-Diagonal Long-Range Order, ODLRO) for superfluidity, and laser light, is related to first-order (1-body) coherence/ODLRO, while superconductivity is related to second-order coherence/ODLRO. (For fermions, such as electrons, only even orders of coherence/ODLRO are possible.) For bosons, a Bose–Einstein condensate is an example of a system exhibiting macroscopic quantum coherence through a multiple occupied single-particle state.

Regarding the occurrence of quantum coherence at a macroscopic level, it is interesting to note that the classical electromagnetic field exhibits macroscopic quantum coherence. The most obvious example is the carrier signal for radio and TV. They satisfy Glauber's quantum description of coherence.

Recently M.B. Plenio and co-workers constructed an operational formulation of Quantum coherence as a resource theory. They introduced coherence monotones analogous to the entanglement monotones.

Spatial Coherence

The Young's double slit experiment essentially measures the spatial coherence. The wave $\tilde{E}(t)$ at the point P on the screen is the superposition of $\tilde{E}_1(t)$ and $\tilde{E}_2(t)$ the contributions from slits 1 and 2 respectively. Let us now shift our attention to the values of the electric field $\tilde{E}_1(t)$and $\tilde{E}_2(t)$ at the positions of the two slits. We define the spatial coherence of the electric field at the two slit positions as

$$C_{12}(d)=\frac{\frac{1}{2}\left\langle \tilde{E}_1(t)\tilde{E}_2^*(t)+\tilde{E}_1^*(t)\tilde{E}_2(t)\right\rangle}{2\sqrt{I_1 I_2}}$$

The waves from the two slits pick up different phases along the path from the slits to the screen. The resulting intensity pattern on the screen can be written as

$$I=I_1+I_2+2\sqrt{I_1 I_2}C_{12}(d)\cos(\phi_2-\phi_1)$$

where $\phi_2 - \phi_1$ is the phase difference in the path from the two slits to the screen. The term $\cos(\phi_2 - \phi_1)$ gives rise to a fringe pattern.

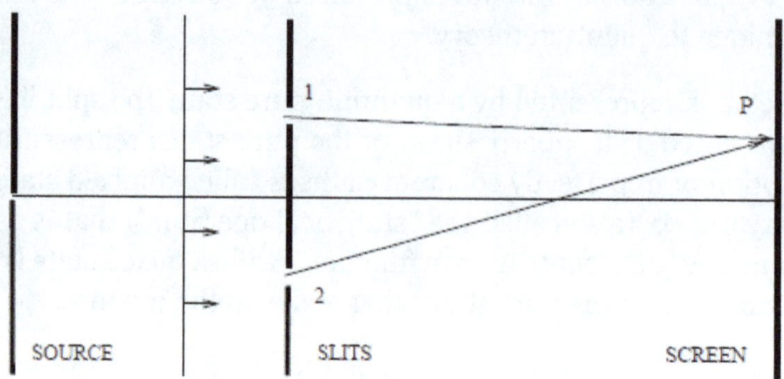

Young's double slit with a point source

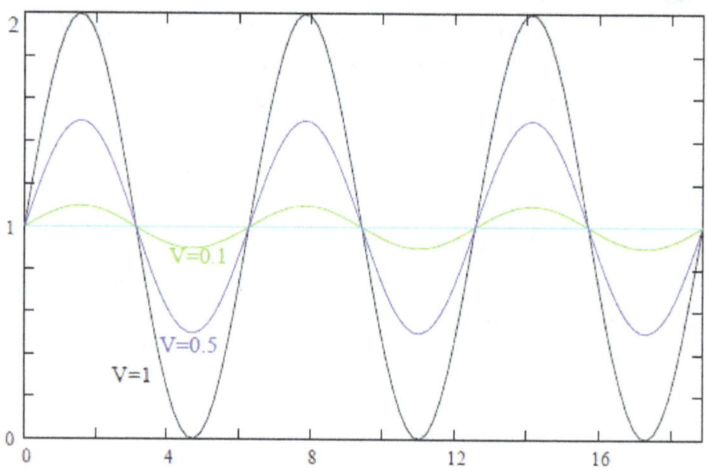

Fringe intensity for different visibilities

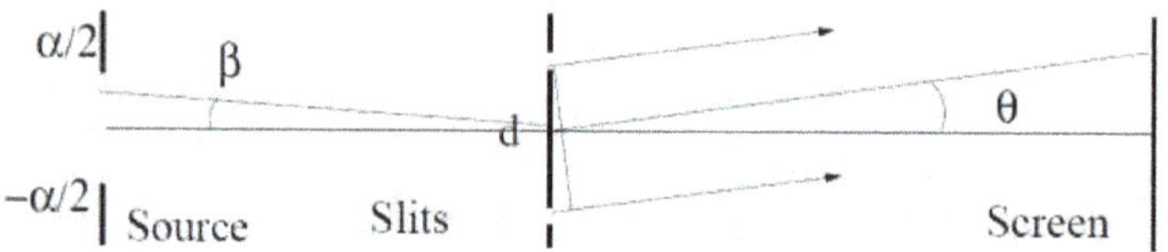

Double slit with a wide source

The fringe visibility defined as

$$V = \frac{I_{max} - I_{min}}{I_{max} + I_{min}}$$

quantifies the contrast of the fringes produced on the screen. It has values in the range $1 \leq V \leq 0$. *A value* $V = 1$ implies very high contrast fringes, the fringes are washed away when $V = 0$. Figure shows the fringe pattern for different values of V. It can be easily checked that the visibility is related to the spatial coherence as

$$V = \frac{2\sqrt{I_1 I_2}\,|C_{12}(d)|}{I_1 + I_2}$$

and the visibility directly gives the spatial coherence $V = |C_{12}(d)|$ when $I_1 = I_2$.

Let us first consider the situation when the two slits are illuminated by a distant point source as shown in Figure. Here the two slits lie on the same wavefront, and $\tilde{E}_1(t) = \tilde{E}_2(t)$. We then have

$$\frac{1}{2}\langle \tilde{E}_1(t)\tilde{E}_2^*(t)\rangle = \frac{1}{2}\langle \tilde{E}_1^*(t)\tilde{E}_2(t)\rangle = I_1 = I_2.$$

whereby $C_{12}(d) = 1$ and the fringes have a visibility $V = 1$.

We next consider the effect of a finite source size. It is assumed that the source subtends an angle α as shown in Figure. This situation can be analyzed by first considering a source at an angle β as shown in the figure.

This produces an intensity

$$I(\theta,\beta) = 2I_1\left[1 + \cos\left(\frac{2\pi d}{\lambda}(\theta + \beta)\right)\right]$$

at a point at an angle θ on the screen where it is assumed that $\theta, \beta \ll 1$. Integrating β over the angular extent of the source

$$\begin{aligned} I(\theta) &= \frac{1}{\alpha}\int_{-\alpha/2}^{\alpha/2} I(\theta,\beta)\,d\beta \\ &= 2I_1\left[1 + \frac{\lambda}{\alpha 2\pi d}\left\{\sin\left[\frac{2\pi d}{\lambda}\left(\theta + \frac{\alpha}{2}\right)\right] - \sin\left[\frac{2\pi d}{\lambda}\left(\theta - \frac{\alpha}{2}\right)\right]\right\}\right] \\ &= 2I_1\left[1 + \frac{\lambda}{\pi d\alpha}\cos\left(\frac{2\pi d\theta}{\lambda}\right)\sin\left(\frac{\pi d\alpha}{\lambda}\right)\right] \end{aligned}$$

It is straightforward to calculate the spatial coherence by comparing eq. This has a value

$$C_{12}(d) = \sin\left(\frac{\pi d\alpha}{\lambda}\right) / \left(\frac{\pi d\alpha}{\lambda}\right)$$

and the visibility is $V = |C_{12}(d)|$. Thus we see that the visibility which quantifies the fringe contrast in the Young's double slit experiment gives a direct estimate of the spatial coherence. The visibility, or equivalently the spatial coherence goes down if the angular extent of the source is increased. It is interesting to note that the visibility becomes exactly zero when the argument of the Sine term the expression becomes integral multiple of π. So when the width of the source is equal to $m\lambda / d, m = 1, 2\cdots$ the visibility is zero.

Why does the fringe contrast go down if the angular extent of the source is increased? This occurs because the two slits are no longer illuminated by a single wavefront, There now are many different wavefronts incident on the slits, one from each point on the source. As a consequence the electric fields at the two slits are no longer perfectly coherent $|C_{12}(d)| < 1$ and the fringe contrast is reduced.

Expression shows how the Young's double slit experiment can be used to determine the angular extent of sources. For example consider a situation where the experiment is done with starlight. The variation of the visibility V or equivalently the spatial coherence $C_{12}(d)$ with varying slit separation d is governed by eq. Measurements of the visibility as a function of d can be used to determine α the angular extent of the star.

Temporal Coherence

The Michelson interferometer measures the temporal coherence of the wave. Here a single wave front $\tilde{E}(t)$ is split into two $\tilde{E}_1(t)$ and $\tilde{E}_2(t)$ at the beam splitter. This is referred to as division of amplitude. The two waves are then superposed, one of the waves being given an extra time delay τ through the difference in the arm lengths. The intensity of the fringes is

$$I = \frac{1}{2}\left\langle \left[\tilde{E}_1(t) + \tilde{E}_2(t+\tau)\right]\left[\tilde{E}_1(t) + \tilde{E}_2(t+\tau)\right]^* \right\rangle$$

$$= I_1 + I_2 + \frac{1}{2}\left\langle \tilde{E}_1(t)\tilde{E}_2^*(t+\tau) + \tilde{E}_1^*(t)\tilde{E}_2(t+\tau)\right\rangle$$

where it is last term involving $\tilde{E}_1(t)\tilde{E}_2^*(t+\tau)$... which is responsible for interference. In our analysis of the Michelson interferometer in the previous chapter we had assumed that the incident wave is purely monochromatic *ie.* $\tilde{E}(t) = \tilde{E}e^{i\omega t}$ whereby

$$\frac{1}{2}\left\langle \tilde{E}_1(t)\tilde{E}_2^*(t+\tau) + \tilde{E}_1^*(t)\tilde{E}_2(t+\tau)\right\rangle = 2\sqrt{I_1 I_2}\cos(\omega\tau)$$

The above assumption is an idealization that we adopt because it simplifies the analysis. In reality we do not have waves of a single frequency, there is always a finite spread in frequencies. How does this affect equation?

As an example let us consider two frequencies $\omega_1 = \omega - \Delta\omega/2$ *and* $\omega_2 = \omega + \Delta\omega/2$ with $\Delta\omega \ll \omega$

$$\tilde{E}(t) = \tilde{a}\left[e^{i\omega_1 t} + e^{i\omega_2 t}\right].$$

This can also be written as

$$\tilde{E}(t) = \tilde{A}(t)e^{i\omega t}$$

which is a wave of angular frequency ω whose amplitude $\tilde{A}(t) = 2\tilde{a}\cos(\Delta\omega t/2)$ varies slowly with time. We now consider a more realistic situation where we have many frequencies in the range $\omega - \Delta\omega/2$ *to* $\omega + \Delta\omega/2$. The resultant will again be of the same form as equation. where there is a wave with angular frequency ω whose amplitude $\tilde{A}(t)$ varies slowly on the timescale

$$T \sim \frac{2\pi}{\Delta\omega}.$$

Note that the amplitude $A(t)$ and phase $\phi(t)$ of the complex amplitude $\tilde{A}(t)$ both vary slowly with timescale T. Figure shows a situation where $\Delta\omega/\omega = 0.2$, a pure sinusoidal wave of the same frequency is shown for comparison. What happens to equation in the presence of a finite spread in frequencies? It now gets modified to

$$\frac{1}{2}\left\langle \tilde{E}_1(t)\tilde{E}_2^*(t+\tau)+\tilde{E}_1^*(t)\tilde{E}_2(t+\tau)\right\rangle = 2\sqrt{I_1 I_2}C_{12}(\tau)\cos(\omega\tau)$$

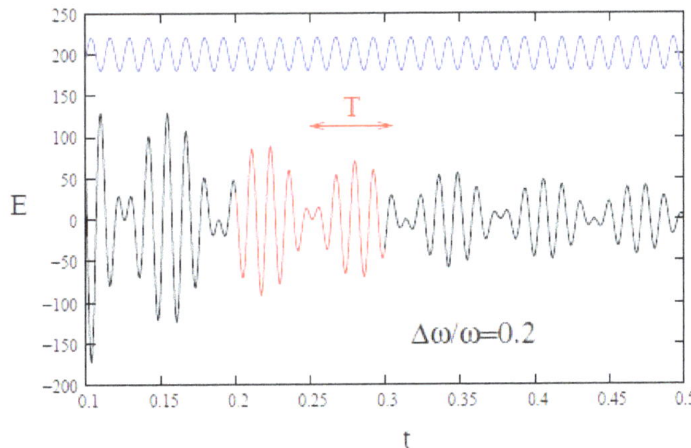

Variation of E with time for monochromatic and polychromaticlight

where $C_{12}(\tau) \le 1$. Here $C_{12}(\tau)$ is the temporal coherence of the two waves $\tilde{E}_1(t)\, and\, \tilde{E}_2(t)$ for a time delay τ. Two waves are perfectly coherent if $C_{12}(\tau)=1$, partially coherent if $0 < C_{12}(\tau) < 1$ and incoherent if $C_{12}(\tau)=0$. Typically the coherence time τ_c of a wave is decided by the spread in frequencies

$$\tau_c = \frac{2\pi}{\Delta\omega}.$$

The waves are coherent for time delays τ less than τ_c $ie.\, C_{12}(\tau) \sim 1\; for\; \tau < \tau_c$, and the waves are incoherent for larger time delays $ie\, C_{12}(\tau) \sim 0\; for\; \tau < \tau_c$.. Interference will be observed only if $\tau < \tau_c$. The coherence time τ_c can be converted to a length-scale $l_c = c\tau_c$ called the coherence length.

An estimate of the frequency spread $\Delta\nu = \Delta\omega/2\pi$ can be made by studying the intensity distribution of a source with respect to frequency. Full width at half maximum (FWHM) of the intensity profile gives a good estimate of the frequency spread.

The Michelson interferometer can be used to measure the temporal coherence $C_{12}(\tau)$. Assuming that $I_1 = I_2$, we have $V = C_{12}(\tau)$. Measuring the visibility of the fringes varying d the difference in the arm lengths of a Michelson interferometer gives an estimate of the temporal coherence for $\tau = d/c$. The fringes will have a good contrast V 1 only for $d < l_c$. The fringes will be washed away for d values larger than l_c.

References

- Pincock, S. (25 February 2011). "Cool laser makes atoms march in time". ABC Science. ABC News Online. Retrieved 2011-03-02
- Albert Michelson; Edward Morley (1887). "On the Relative Motion of the Earth and the Luminiferous Ether". American Journal of Science. 34 (203): 333–345. doi:10.2475/ajs.s3-34.203.333
- Malacara, D. (2007). "Twyman–Green Interferometer". Optical Shop Testing. p. 46. doi:10.1002/9780470135976.ch2. ISBN 9780470135976
- Gary, G.A.; Balasubramaniam, K.S. "Additional Notes Concerning the Selection of a Multiple-Etalon System for ATST" (PDF). Advanced Technology Solar Telescope. Retrieved 29 April 2012
- Michelson, A.A. (1881). "The Relative Motion of the Earth and the Luminiferous Ether". American Journal of Science. 22: 120–129. doi:10.2475/ajs.s3-22.128.120
- Leonard Mandel; Emil Wolf (1995). Optical Coherence and Quantum Optics. Cambridge University Press. ISBN 0-521-41711-2
- Olszak, A.G.; Schmit, J.; Heaton, M.G. "Interferometry: Technology and Applications" (PDF). Bruker. Retrieved 1 April 2012
- Shankland, R.S. (1964). "Michelson–Morley experiment". American Journal of Physics. 31 (1): 16–35. Bibcode:1964AmJPh..32...16S. doi:10.1119/1.1970063
- Christopher Gerry; Peter Knight (2005). Introductory Quantum Optics. Cambridge University Press. ISBN 978-0-521-52735-4
- Dean Pesnell; Kevin Addison (5 February 2010). "SDO - Solar Dynamics Observatory: SDO Instruments". NASA. Retrieved 2010-02-13
- Fercher, A.F. (1996). "Optical Coherence Tomography" (PDF). Journal of Biomedical Optics. 1 (2): 157–173. Bibcode:1996JBO.....1..157F. doi:10.1117/12.231361. Retrieved 10 April 2012
- Solar Physics Research Group. "Helioseismic and Magnetic Imager Investigation". Stanford University. Retrieved 2010-02-13

5

Diffraction: An Overview

The phenomenon where light bends due to simultaneous interference of many waves is known as diffraction. Diffraction occurs in all types of waves. It is very important in the creation of the X-ray. The topics discussed in the chapter are of great importance to broaden the existing knowledge on diffraction.

Diffraction

Diffraction refers to various phenomena that occur when a wave encounters an obstacle or a slit. It is defined as the bending of light around the corners of an obstacle or aperture into the region of geometrical shadow of the obstacle. In classical physics, the diffraction phenomenon is described as the interference of waves according to the Huygens–Fresnel principle. These characteristic behaviors are exhibited when a wave encounters an obstacle or a slit that is comparable in size to its wavelength. Similar effects occur when a light wave travels through a medium with a varying refractive index, or when a sound wave travels through a medium with varying acoustic impedance. Diffraction occurs with all waves, including sound waves, water waves, and electromagnetic waves such as visible light, X-rays and radio waves.

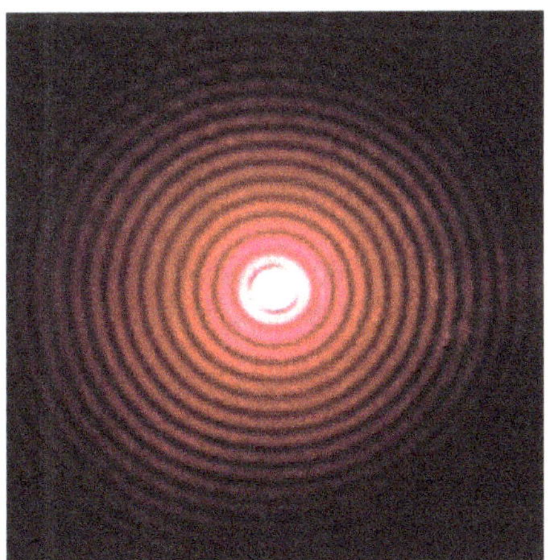

Diffraction pattern of red laser beam made on a plate after passing a small circular hole in another plate

Since physical objects have wave-like properties (at the atomic level), diffraction also occurs with matter and can be studied according to the principles of quantum mechanics. Italian scientist Francesco Maria Grimaldi coined the word "diffraction" and was the first to record accurate observations of the phenomenon in 1660.

While diffraction occurs whenever propagating waves encounter such changes, its effects are generally most pronounced for waves whose wavelength is roughly comparable to the dimensions of the diffracting object or slit. If the obstructing object provides multiple, closely spaced openings, a complex pattern of varying intensity can result. This is due to the addition, or interference, of different parts of a wave that travel to the observer by different paths, where different path lengths result in different phases. The formalism of diffraction can also describe the way in which waves of finite extent propagate in free space. For example, the expanding profile of a laser beam, the beam shape of a radar antenna and the field of view of an ultrasonic transducer can all be analyzed using diffraction equations.

Examples

Solar glory at the steam from hot springs. A glory is an optical phenomenon produced by light backscattered (a combination of diffraction, reflection and refraction) towards its source by a cloud of uniformly sized water droplets

The effects of diffraction are often seen in everyday life. The most striking examples of diffraction are those that involve light; for example, the closely spaced tracks on a CD or DVD act as a diffraction grating to form the familiar rainbow pattern seen when looking at a disc. This principle can be extended to engineer a grating with a structure such that it will produce any diffraction pattern desired; the hologram on a credit card is an example. Diffraction in the atmosphere by small particles can cause a bright ring to be visible around a bright light source like the sun or the moon. A shadow of a solid object, using light from a compact source, shows small fringes near its edges. The speckle pattern which is observed when laser light falls on an optically rough surface is also a diffraction phenomenon. When deli meat appears to be iridescent, that is diffraction off the meat fibers. All these effects are a consequence of the fact that light propagates as a wave.

Diffraction can occur with any kind of wave. Ocean waves diffract around jetties and other obstacles. Sound waves can diffract around objects, which is why one can still hear someone calling even when hiding behind a tree. Diffraction can also be a concern in some technical applications; it sets a fundamental limit to the resolution of a camera, telescope, or microscope.

History

The law of refraction (not diffraction) of light, which would much later come to be known as Snell's law was first accurately described by the scientist Ibn Sahl at the Baghdad court in 984. In the manuscript *On Burning Mirrors and Lenses*, Sahl used the law to derive lens shapes that focus light with no geometric aberrations.

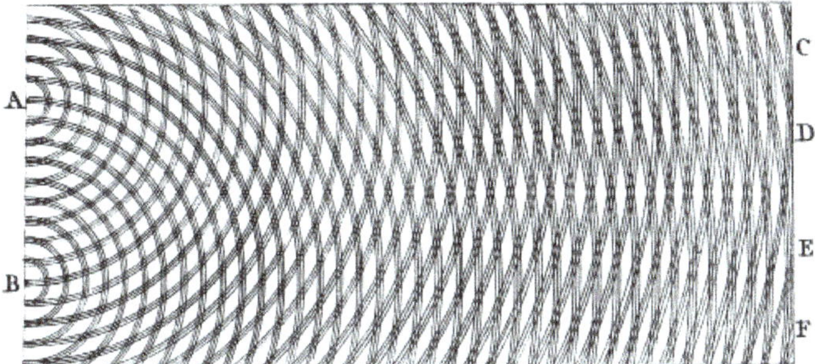

Thomas Young's sketch of two-slit diffraction, which he presented to the Royal Society in 1803.

The effects of diffraction of light were first carefully observed and characterized by Francesco Maria Grimaldi, who also coined the term *diffraction*, from the Latin *diffringere*, 'to break into pieces', referring to light breaking up into different directions. The results of Grimaldi's observations were published posthumously in 1665. Isaac Newton studied these effects and attributed them to *inflexion* of light rays. James Gregory (1638–1675) observed the diffraction patterns caused by a bird feather, which was effectively the first diffraction grating to be discovered. Thomas Young performed a celebrated experiment in 1803 demonstrating interference from two closely spaced slits. Explaining his results by interference of the waves emanating from the two different slits, he deduced that light must propagate as waves. Augustin-Jean Fresnel did more definitive studies and calculations of diffraction, made public in 1815 and 1818, and thereby gave great support to the wave theory of light that had been advanced by Christiaan Huygens and reinvigorated by Young, against Newton's particle theory.

Mechanism

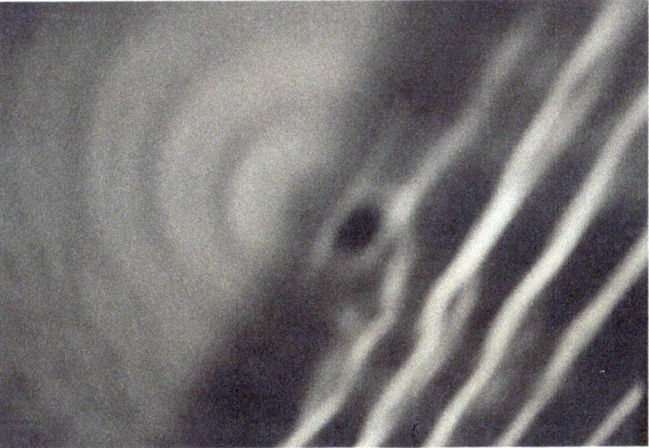
Photograph of single-slit diffraction in a circular ripple tank

In traditional classical physics diffraction arises because of the way in which waves propagate; this is described by the Huygens–Fresnel principle and the principle of superposition of waves. The propagation of a wave can be visualized by considering every particle of the transmitted medium on a wavefront as a point source for a secondary spherical wave. The wave displacement at any subsequent point is the sum of these secondary waves. When waves are added together, their sum is determined by the relative phases as well as the amplitudes of the individual waves so that the summed amplitude of the waves can have any value between zero and the sum of the individual amplitudes. Hence, diffraction patterns usually have a series of maxima and minima.

In the modern quantum mechanical understanding of light propagation through a slit (or slits) every photon has what is known as a wavefunction which describes its path from the emitter through the slit to the screen. The wavefunction (the path the photon will take) is determined by the physical surroundings such as slit geometry, screen distance and initial conditions when the photon is created. In important experiments (A low-intensity double-slit experiment was first performed by G. I. Taylor in 1909) the existence of the photon's wavefunction was demonstrated. In the quantum approach the diffraction pattern is created by the distribution of paths, the observation of light and dark bands is the presence or absence of photons in these areas (no interference!). The quantum approach has some striking similarities to the Huygens-Fresnel principle, in that principle the light becomes a series of individually distributed light sources across the slit which is similar to the limited number of paths (or wave functions) available for the photons to travel through the slit.

There are various analytical models which allow the diffracted field to be calculated, including the Kirchhoff-Fresnel diffraction equation which is derived from wave equation, the Fraunhofer diffraction approximation of the Kirchhoff equation which applies to the far field and the Fresnel diffraction approximation which applies to the near field. Most configurations cannot be solved analytically, but can yield numerical solutions through finite element and boundary element methods.

It is possible to obtain a qualitative understanding of many diffraction phenomena by considering how the relative phases of the individual secondary wave sources vary, and in particular, the conditions in which the phase difference equals half a cycle in which case waves will cancel one another out.

The simplest descriptions of diffraction are those in which the situation can be reduced to a two-dimensional problem. For water waves, this is already the case; water waves propagate only on the surface of the water. For light, we can often neglect one direction if the diffracting object extends in that direction over a distance far greater than the wavelength. In the case of light shining through small circular holes we will have to take into account the full three-dimensional nature of the problem.

Diffraction of Light

Some examples of diffraction of light are considered below.

Single-slit Diffraction

Numerical approximation of diffraction pattern from a slit of width equal to wavelength of an incident plane wave in 3D spectrum visualization

Numerical approximation of diffraction pattern from a slit of width equal to five times the wavelength of an incident plane wave in 3D spectrum visualization

A long slit of infinitesimal width which is illuminated by light diffracts the light into a series of circular waves and the wavefront which emerges from the slit is a cylindrical wave of uniform intensity.

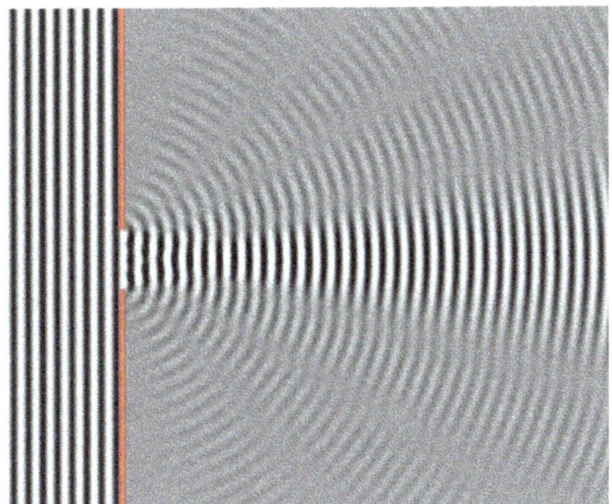

Numerical approximation of diffraction pattern from a slit of width four wavelengths with an incident plane wave. The main central beam, nulls, and phase reversals are apparent

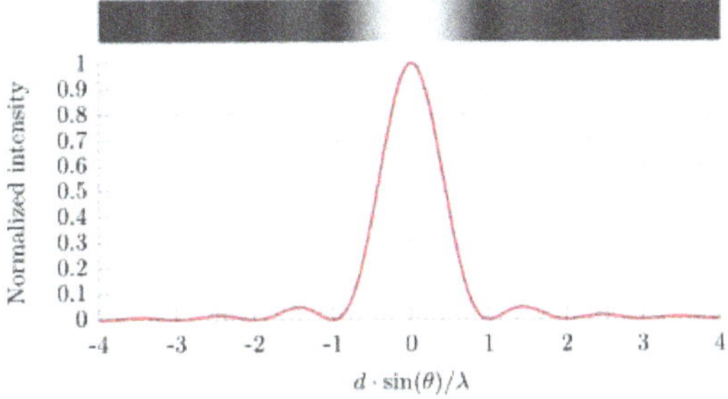

Graph and image of single-slit diffraction

A slit which is wider than a wavelength produces interference effects in the space downstream of the slit. These can be explained by assuming that the slit behaves as though it has a large number of point sources spaced evenly across the width of the slit. The analysis of this system is simplified if we consider light of a single wavelength. If the incident light is coherent, these sources all have the same phase. Light incident at a given point in the space downstream of the slit is made up of contributions from each of these point sources and if the relative phases of these contributions vary by 2π or more, we may expect to find minima and maxima in the diffracted light. Such phase differences are caused by differences in the path lengths over which contributing rays reach the point from the slit.

We can find the angle at which a first minimum is obtained in the diffracted light by the following reasoning. The light from a source located at the top edge of the slit interferes destructively with a source located at the middle of the slit, when the path difference between them is equal to $\lambda/2$. Similarly, the source just below the top of the slit will interfere destructively with the source located just below the middle of the slit at the same angle. We can continue this reasoning along the entire height of the slit to conclude that the condition for destructive interference for the entire slit is the same as the condition for destructive interference between two narrow slits a distance apart that is half the width of the slit. The path difference is given by

$\frac{d\sin(\theta)}{2}$ so that the minimum intensity occurs at an angle $\theta_{\min}$ given by

$$d\sin\theta_{\min} = \lambda$$

where

- d is the width of the slit,
- $\theta_{\min}$ is the angle of incidence at which the minimum intensity occurs, and
- λ is the wavelength of the light

A similar argument can be used to show that if we imagine the slit to be divided into four, six, eight parts, etc., minima are obtained at angles θ_n given by

$$d\sin\theta_n = n\lambda$$

where

- n is an integer other than zero.

There is no such simple argument to enable us to find the maxima of the diffraction pattern. The intensity profile can be calculated using the Fraunhofer diffraction equation as

$$I(\theta) = I_0 \operatorname{sinc}^2\left(\frac{d\pi}{\lambda}\sin\theta\right)$$

where

- $I(\theta)$ is the intensity at a given angle,
- I_0 is the original intensity, and
- the unnormalized sinc function above is given by $sinc(x) = \frac{\sin x}{x}$ if $x \neq 0$, and $sinc(0) = 1$

This analysis applies only to the far field, that is, at a distance much larger than the width of the slit.

Diffraction Grating

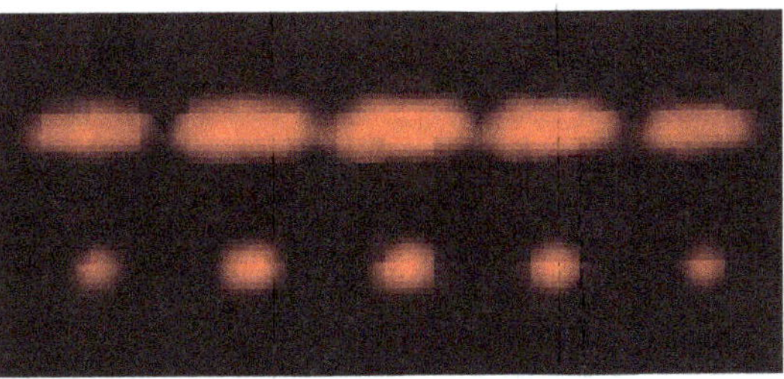

2-slit (top) and 5-slit diffraction of red laser light

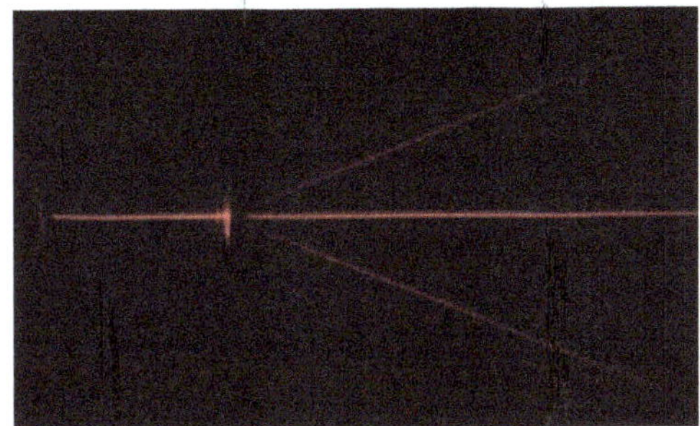

Diffraction of a red laser using a diffraction grating.

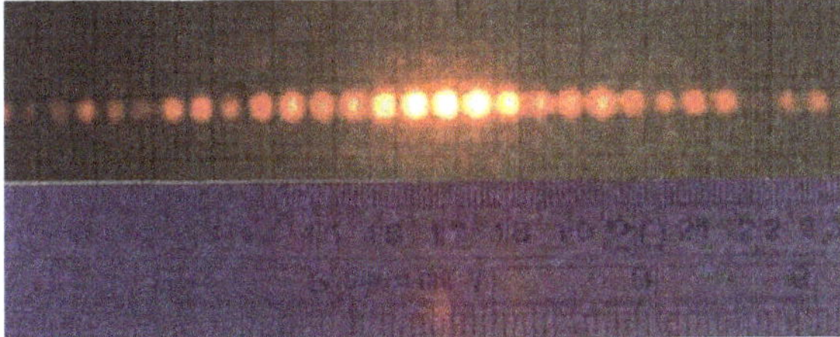

A diffraction pattern of a 633 nm laser through a grid of 150 slits

A diffraction grating is an optical component with a regular pattern. The form of the light diffracted by a grating depends on the structure of the elements and the number of elements present, but all gratings have intensity maxima at angles θ_m which are given by the grating equation

$$d(\sin\theta_m + \sin\theta_i) = m\lambda.$$

where

- θ_i is the angle at which the light is incident,
- d is the separation of grating elements, and
- m is an integer which can be positive or negative.

The light diffracted by a grating is found by summing the light diffracted from each of the elements, and is essentially a convolution of diffraction and interference patterns.

The figure shows the light diffracted by 2-element and 5-element gratings where the grating spacings are the same; it can be seen that the maxima are in the same position, but the detailed structures of the intensities are different.

Circular Aperture

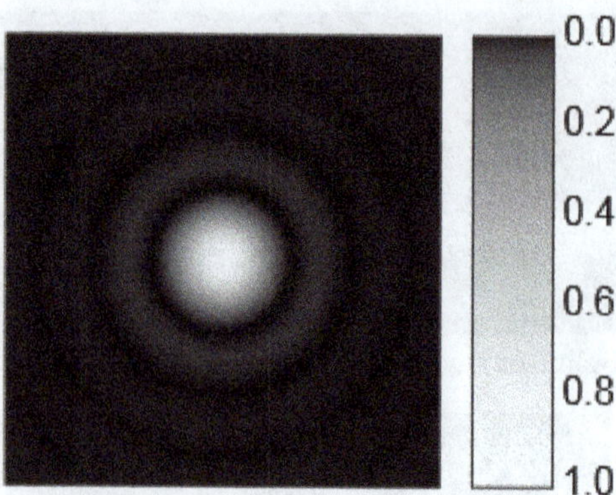

A computer-generated image of an Airy disk

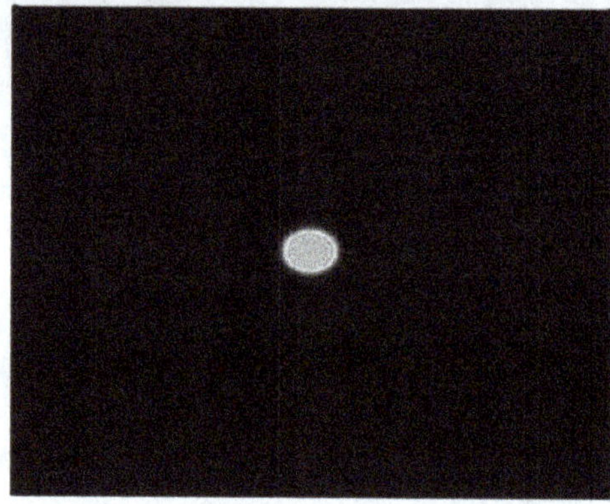

Computer generated light diffraction pattern from a circular aperture of diameter 0.5 micrometre at a wavelength of 0.6 micrometre (red-light) at distances of 0.1 cm – 1 cm in steps of 0.1 cm.

The far-field diffraction of a plane wave incident on a circular aperture is often referred to as the Airy Disk. The variation in intensity with angle is given by

$$I(\theta) = I_0 \left(\frac{2J_1(ka\sin\theta)}{ka\sin\theta} \right)^2,$$

where a is the radius of the circular aperture, k is equal to $2\pi/\lambda$ and J_1 is a Bessel function. The smaller the aperture, the larger the spot size at a given distance, and the greater the divergence of the diffracted beams.

General Aperture

The wave that emerges from a point source has amplitude ψ at location r that is given by the solution of the frequency domain wave equation for a point source (The Helmholtz Equation),

$$\nabla^2\psi + k^2\psi = \delta(\mathbf{r})$$

where $\delta(\mathbf{r})$ is the 3-dimensional delta function. The delta function has only radial dependence, so the Laplace operator (a.k.a. scalar Laplacian) in the spherical coordinate system simplifies to

$$\nabla^2 \psi = \frac{1}{r}\frac{\partial^2}{\partial r^2}(r\psi)$$

By direct substitution, the solution to this equation can be readily shown to be the scalar Green's function, which in the spherical coordinate system (and using the physics time convention $e^{-i\omega t}$) is:

$$\psi(r) = \frac{e^{ikr}}{4\pi r}$$

This solution assumes that the delta function source is located at the origin. If the source is located at an arbitrary source point, denoted by the vector $\mathbf{r}'$ and the field point is located at the point $\mathbf{r}$, then we may represent the scalar Green's function (for arbitrary source location) as:

$$\psi(\mathbf{r}\,|\,\mathbf{r}') = \frac{e^{ik|\mathbf{r}-\mathbf{r}'|}}{4\pi\,|\mathbf{r}-\mathbf{r}'|}$$

Therefore, if an electric field, $E_{inc}(x,y)$ is incident on the aperture, the field produced by this aperture distribution is given by the surface integral:

$$\Psi(r) \propto \iint_{\text{aperture}} E_{\text{inc}}(x',y')\frac{e^{ik|\mathbf{r}-\mathbf{r}'|}}{4\pi\,|\mathbf{r}-\mathbf{r}'|}dx'dy',$$

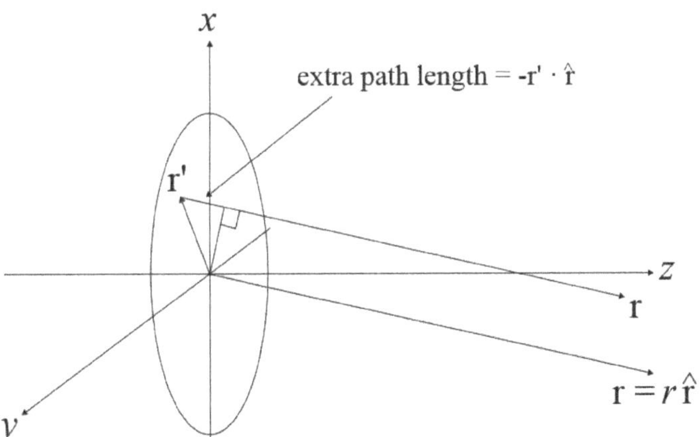

On the calculation of Fraunhofer region fields

where the source point in the aperture is given by the vector

$$\mathbf{r}' = x'\hat{\mathbf{x}} + y'\hat{\mathbf{y}}$$

In the far field, wherein the parallel rays approximation can be employed, the Green's function,

$$\psi(\mathbf{r}\,|\,\mathbf{r}') = \frac{e^{ik|\mathbf{r}-\mathbf{r}'|}}{4\pi\,|\mathbf{r}-\mathbf{r}'|}$$

simplifies to

$$\psi(\mathbf{r}\,|\,\mathbf{r}') = \frac{e^{ikr}}{4\pi r} e^{-ik(\mathbf{r}'\cdot\hat{\mathbf{r}})}$$

as can be seen in the figure to the right.

The expression for the far-zone (Fraunhofer region) field becomes

$$\Psi(r) \propto \frac{e^{ikr}}{4\pi r} \iint_{\text{aperture}} E_{\text{inc}}(x', y') e^{-ik(\mathbf{r}'\cdot\hat{\mathbf{r}})} dx'dy',$$

Now, since

$$\mathbf{r}' = x'\hat{\mathbf{x}} + y'\hat{\mathbf{y}}$$

and

$$\hat{\mathbf{r}} = \sin\theta\cos\phi\hat{\mathbf{x}} + \sin\theta\,\sin\phi\hat{\mathbf{y}} + \cos\theta\hat{\mathbf{z}}$$

the expression for the Fraunhofer region field from a planar aperture now becomes,

$$\Psi(r) \propto \frac{e^{ikr}}{4\pi r} \iint_{\text{aperture}} E_{\text{inc}}(x', y') e^{-ik\sin\theta(\cos\phi x' + \sin\phi y')} dx'dy'$$

Letting,

$$k_x = k\sin\theta\cos\phi$$

and

$$k_y = k\sin\theta\sin\phi$$

the Fraunhofer region field of the planar aperture assumes the form of a Fourier transform

$$\Psi(r) \propto \frac{e^{ikr}}{4\pi r} \iint_{\text{aperture}} E_{\text{inc}}(x', y') e^{-i(k_x x' + k_y y')} dx'dy',$$

In the far-field / Fraunhofer region, this becomes the spatial Fourier transform of the aperture distribution. Huygens' principle when applied to an aperture simply says that the far-field diffraction pattern is the spatial Fourier transform of the aperture shape, and this is a direct by-product of using the parallel-rays approximation, which is identical to doing a plane wave decomposition of the aperture plane fields.

Propagation of a Laser Beam

The way in which the beam profile of a laser beam changes as it propagates is determined by diffraction. When the entire emitted beam has a planar, spatially coherent wave front, it approximates Gaussian beam profile and has the lowest divergence for a given diameter. The smaller the output beam, the quicker it diverges. It is possible to reduce the divergence of a laser beam by first expanding it with one convex lens, and then collimating it with a second convex lens whose focal point is coincident with that of the first lens. The resulting beam has a larger diameter, and hence a lower divergence. Divergence of a laser beam may be reduced below the diffraction of a Gaussian beam or even reversed to convergence if the refractive index of the propagation media increases with the light intensity. This may result in a self-focusing effect.

When the wave front of the emitted beam has perturbations, only the transverse coherence length (where the wave front perturbation is less than 1/4 of the wavelength) should be considered as a Gaussian beam diameter when determining the divergence of the laser beam. If the transverse coherence length in the vertical direction is higher than in horizontal, the laser beam divergence will be lower in the vertical direction than in the horizontal.

Diffraction-limited Imaging

The ability of an imaging system to resolve detail is ultimately limited by diffraction. This is because a plane wave incident on a circular lens or mirror is diffracted as described above. The light is not focused to a point but forms an Airy disk having a central spot in the focal plane with radius to first null of

$$d = 1.22\lambda N,$$

where λ is the wavelength of the light and N is the f-number (focal length divided by diameter) of the imaging optics. In object space, the corresponding angular resolution is

$$\sin\theta = 1.22\frac{\lambda}{D},$$

where D is the diameter of the entrance pupil of the imaging lens (e.g., of a telescope's main mirror).

Two point sources will each produce an Airy pattern – see the photo of a binary star. As the point sources move closer together, the patterns will start to overlap, and ultimately they will merge to form a single pattern, in which case the two point sources cannot be resolved in the image. The Rayleigh criterion specifies that two point sources can be considered to be resolvable if the separation of the two images is at least the radius of the Airy disk, i.e. if the first minimum of one coincides with the maximum of the other.

Thus, the larger the aperture of the lens, and the smaller the wavelength, the finer the resolution of an imaging system. This is why telescopes have very large lenses or mirrors, and why optical microscopes are limited in the detail which they can see.

Speckle Patterns

The speckle pattern which is seen when using a laser pointer is another diffraction phenomenon. It is a result of the superposition of many waves with different phases, which are produced when a laser beam illuminates a rough surface. They add together to give a resultant wave whose amplitude, and therefore intensity, varies randomly.

Babinet's Principle

Babinet's Principle is a useful theorem stating that the diffraction pattern from an opaque body is identical to that from a hole of the same size and shape, but with differing intensities. This means that the interference conditions of a single obstruction would be the same as that of a single slit.

Patterns

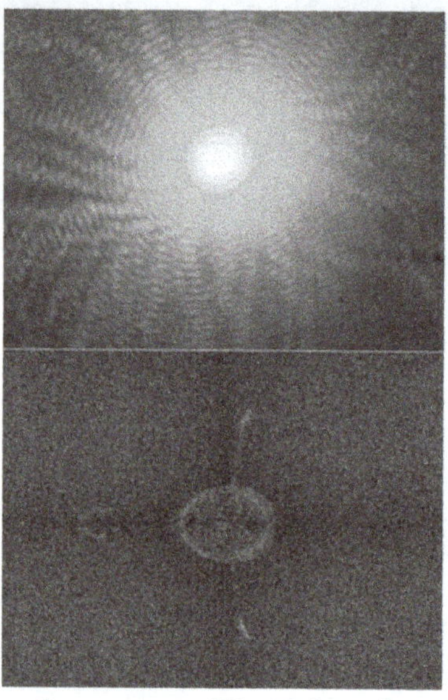

The upper half of this image shows a diffraction pattern of He-Ne laser beam on an elliptic aperture. The lower half is its 2D Fourier transform approximately reconstructing the shape of the aperture

Several qualitative observations can be made of diffraction in general:

- The angular spacing of the features in the diffraction pattern is inversely proportional to the dimensions of the object causing the diffraction. In other words: The smaller the diffracting object, the 'wider' the resulting diffraction pattern, and vice versa. (More precisely, this is true of the sines of the angles.)
- The diffraction angles are invariant under scaling; that is, they depend only on the ratio of the wavelength to the size of the diffracting object.
- When the diffracting object has a periodic structure, for example in a diffraction grating, the features generally become sharper. The third figure, for example, shows a comparison of a double-slit pattern with a pattern formed by five slits, both sets of slits having the same spacing, between the center of one slit and the next.

Particle Diffraction

Quantum theory tells us that every particle exhibits wave properties. In particular, massive particles can interfere and therefore diffract. Diffraction of electrons and neutrons stood as one of the powerful arguments in favor of quantum mechanics. The wavelength associated with a particle is the de Broglie wavelength

$$\lambda = \frac{h}{p}$$

where h is Planck's constant and p is the momentum of the particle (mass × velocity for slow-moving particles).

For most macroscopic objects, this wavelength is so short that it is not meaningful to assign a wavelength to them. A sodium atom traveling at about 30,000 m/s would have a De Broglie wavelength of about 50 pico meters.

Because the wavelength for even the smallest of macroscopic objects is extremely small, diffraction of matter waves is only visible for small particles, like electrons, neutrons, atoms and small molecules. The short wavelength of these matter waves makes them ideally suited to study the atomic crystal structure of solids and large molecules like proteins.

Relatively larger molecules like buckyballs were also shown to diffract.

Bragg Diffraction

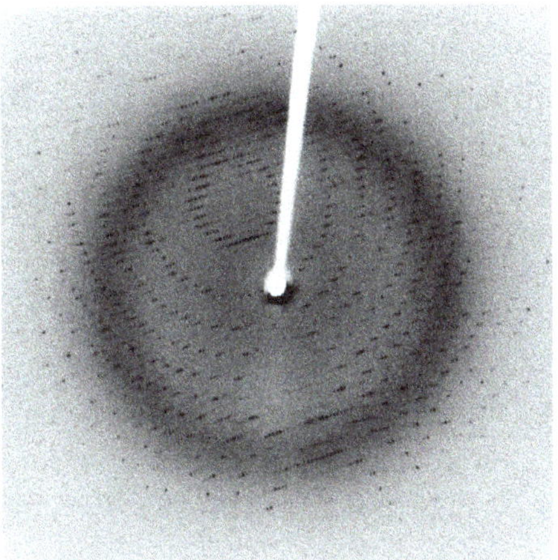

Following Bragg's law, each dot (or *reflection*) in this diffraction pattern forms from the constructive interference of X-rays passing through a crystal. The data can be used to determine the crystal's atomic structure

Diffraction from a three-dimensional periodic structure such as atoms in a crystal is called Bragg diffraction. It is similar to what occurs when waves are scattered from a diffraction grating. Bragg diffraction is a consequence of interference between waves reflecting from different crystal planes. The condition of constructive interference is given by *Bragg's law*:

$$m\lambda = 2d\sin\theta$$

where

λ is the wavelength,

d is the distance between crystal planes,

θ is the angle of the diffracted wave.

and m is an integer known as the *order* of the diffracted beam.

Bragg diffraction may be carried out using either light of very short wavelength like X-rays or matter waves like neutrons (and electrons) whose wavelength is on the order of (or much smaller than) the atomic spacing. The pattern produced gives information of the separations of crystallographic planes d, allowing one to deduce the crystal structure. Diffraction contrast, in electron microscopes and x-topography devices in particular, is also a powerful tool for examining individual defects and local strain fields in crystals.

Coherence

The description of diffraction relies on the interference of waves emanating from the same source taking different paths to the same point on a screen. In this description, the difference in phase between waves that took different paths is only dependent on the effective path length. This does not take into account the fact that waves that arrive at the screen at the same time were emitted by the source at different times. The initial phase with which the source emits waves can change over time in an unpredictable way. This means that waves emitted by the source at times that are too far apart can no longer form a constant interference pattern since the relation between their phases is no longer time independent.

The length over which the phase in a beam of light is correlated, is called the coherence length. In order for interference to occur, the path length difference must be smaller than the coherence length. This is sometimes referred to as spectral coherence, as it is related to the presence of different frequency components in the wave. In the case of light emitted by an atomic transition, the coherence length is related to the lifetime of the excited state from which the atom made its transition.

If waves are emitted from an extended source, this can lead to incoherence in the transversal direction. When looking at a cross section of a beam of light, the length over which the phase is correlated is called the transverse coherence length. In the case of Young's double slit experiment, this would mean that if the transverse coherence length is smaller than the spacing between the two slits, the resulting pattern on a screen would look like two single slit diffraction patterns.

In the case of particles like electrons, neutrons and atoms, the coherence length is related to the spatial extent of the wave function that describes the particle.

Our discussion of interference in the previous chapter considered the superposition of two waves. The discussion can be generalized to a situation where there are three or even more waves. Diffraction is essentially the same as interference, except that we have a superposition of a very large number of waves. In some situations the number is infinite. An obstruction placed along the path of a wave, is an example of a situation where diffraction occurs. In the case of diffraction the longer wavelengths deviate more compared to the shorter wavelengths, whereas in refraction the shorter wavelengths deviate more compared to the longer ones. In the case of diffraction it is necessary to solve the equation governing the wave in the complicated geometry produced by the obstruction. This is usually very complicated, and beyond analytic treatment in most situations. The Huygens-Fresnel principle is a heuristic approach which allows such problems to be handled with relative ease.

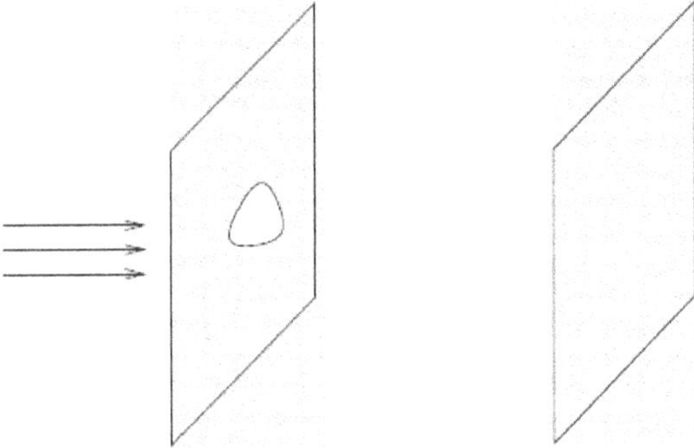

Diffraction due to an obstruction

Huygens' principle, proposed around 1680, states that every point on a wavefront acts as the source for secondary spherical wavelets such that the wavefront at a later time is the envelope of these wavelets. Further, the frequency and propagation speed of the wavelets is the same as that of the wave. Applying Huygens' principle to the propagation of a plane wave, we see that a plane wave front evolves into a shifted plane wave at a later time.

Wavefronts and Huygens' secondary wavelets

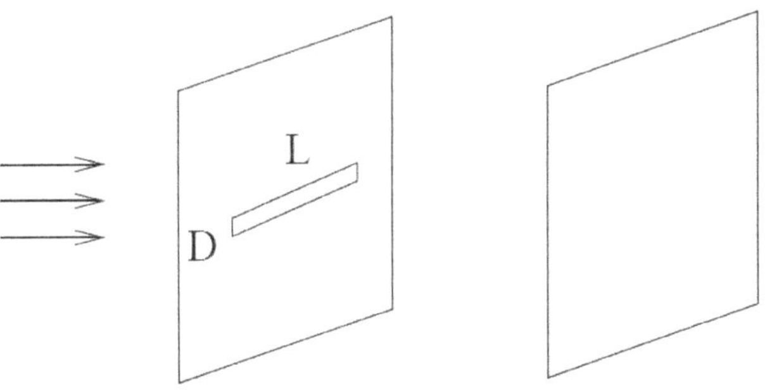

Single slit diffraction

We also show how it can be applied to the propagation of a spherical wavefront. It is also useful for studying the propagation of waves through a refracting medium where the light speed changes with position and is also different in different directions.

Huygens' principle, in its original form cannot be used to explain interference or diffraction. It was modified in the early 19th century by Fresnel to explain diffraction. The modified version is referred to as the Huygens-Fresnel principle. Kirchoff later showed that the Huygens-Fresnel principle is actually consistent with the wave equation that governs the propagation of light.

The Huygens-Fresnel principle states that every unobstructed point on the wavefront acts like a source for a secondary wavelet. The contribution from all these wavelets are to be superposed to find the resultant at any point.

Single Slit Diffraction Pattern

Consider a situation where light from a distant source falls on a rectangular slit of width D and length L as shown in Figure. We shall assume that L is much larger than D. We are interested in the image on a screen which is at a great distance from the slit.

As the source is very far away, we can treat the incident light as plane wave. Further the source is aligned so that the incident wavefronts are parallel to the plane of the slit. Every point in the slit emits a spherical secondary wavelet.These secondary wavelets are well described by plane waves by the time they reach the distant screen. This situation where the incident wave and the emergent secondary waves can all be treated as plane waves is referred to as Fraunhofer diffraction. In this case both, the source as well as the screen are effectively at infinity from the obstacle.

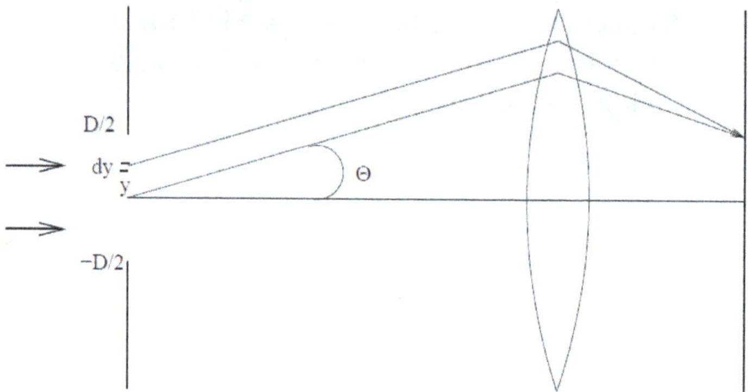

Single slit effective one dimensional arrangement

We assume L to be very large so that the problem can be treated as one dimensional as shown in Figure. Instead of placing the screen far away, it is equivalent to introduce a lens and place the screen at the focal plane. Each point on the slit acts like a secondary source. These secondary sources are all in phase and they all emit secondary wavelets with the same phase. Let us calculate the total radiation at a point at an angle θ. If

$$d\tilde{E} = \tilde{A}\ dy$$

be the contribution from a small element dy at the center of the slit, the contribution from an element a distance y away will be at a different phase

$$d\tilde{E} = \tilde{A}\ e^{i\delta} dy$$

where

$$\delta = \frac{2\pi}{\lambda} y \sin\theta = ky$$

The total electric field can be calculated by adding up the contribution from all points on the slit. This is an integral

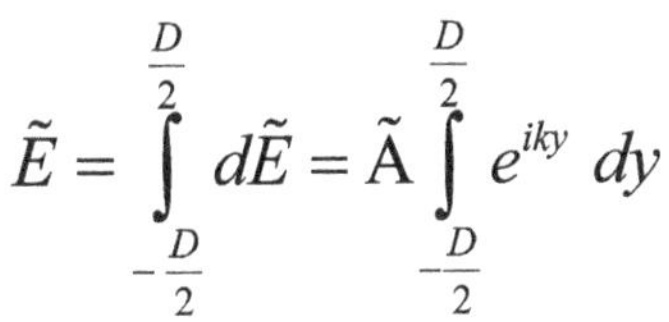

$$\tilde{E} = \int_{-\frac{D}{2}}^{\frac{D}{2}} d\tilde{E} = \tilde{A} \int_{-\frac{D}{2}}^{\frac{D}{2}} e^{iky}\, dy$$

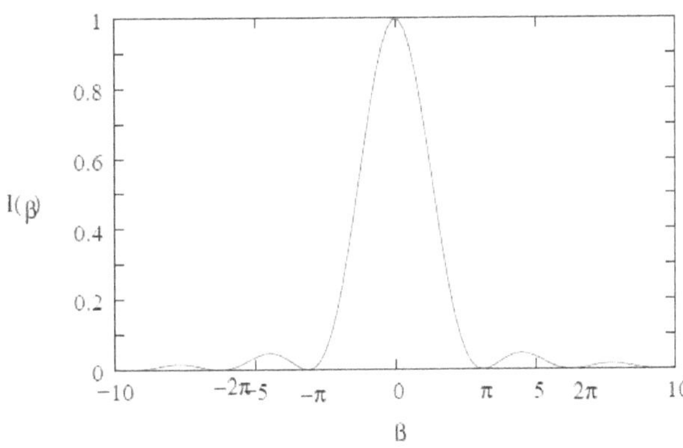

Single slit intensity pattern

$$= \tilde{A} D \sin\left(\frac{kD}{2}\right)\left(\frac{kD}{2}\right)^{-1} = \tilde{A} D \operatorname{sinc}\left(\frac{kD}{2}\right)$$

where $sinc(x) = sin(x)/x$. We use eq. (18.4) to calculate the intensity of light at an angle θ. Defining

$$\beta = \frac{\pi D \sin\theta}{\lambda},$$

the intensity is given by

$$I(\beta) = \frac{1}{2} EE^* = I_0 \sin c^2 \beta.$$

Figure shows the intensity as a function of β. The intensity is maximum at the center where $\beta = 0$. In addition to the oscillations, the intensity falls off proportional to β^2 away from the center. The analysis of the intensity pattern is further simplified In the situation where $\theta \ll 1$ as

$$\beta = \frac{\pi D \theta}{\lambda}.$$

The zeros of the intensity pattern occur at $\beta = \pm m\pi \quad (m, = 1, 2, 3, \ldots)$, or in terms of the angle θ, the zeros are at

$$\theta = m\frac{\lambda}{D}$$

There are intensity maxima located between the zeros. The central maximum at $\theta = 0$ is the brightest, and its angular separation from the nearest zero is λ / D. This gives an estimate of the angular width of the central maximum. The intensities of the other maxima fall away from the center.

Let us compare the intensity pattern $I(\beta)$ shown in Figure with the Predictions of geometrical optics. Figure shows a beam of parallel rays incident on a slit of dimension D. In geometrical optics the only effect of the slit is to cut off some of the rays in the incident beam and reduce the transverse extent of the beam. We expect a beam of parallel rays with transverse dimension D to emerge from the slit. This beam is now incident on a lens which will focuses all the rays to a single point on the screen. Thus in geometrical optics the image is a single bright point on the screen. In reality the wave nature of light manifests itself through the phenomena of diffraction, and we see a pattern of bright spots as shown in Figure. The central spot is the brightest and it has an angular extent $\pm\lambda / D$ The other spots located above and below the central spot are fainter. Taking into account both dimensions of the slit we have

$$I(\beta_x, \beta_y) = I_0 \sin c^2(\beta_x) \sin c^2(\beta_y)$$

Where

$$\beta_x = \frac{\pi L \theta_x}{\lambda} \quad \text{and} \quad \beta_y = \frac{\pi D \theta_y}{\lambda}.$$

and θ_x *and* θ_y are the angles along the x and y axis respectively. The diffraction effects are important on angular scales $\theta_x \sim \lambda / L$ *and* $\theta_y \sim \lambda / D$. In the situation where $L \gg D$ the diffraction effects along θ_x will not be discernable, and we can treat it as a one dimensional slit of dimension D.

Diffraction Formalism

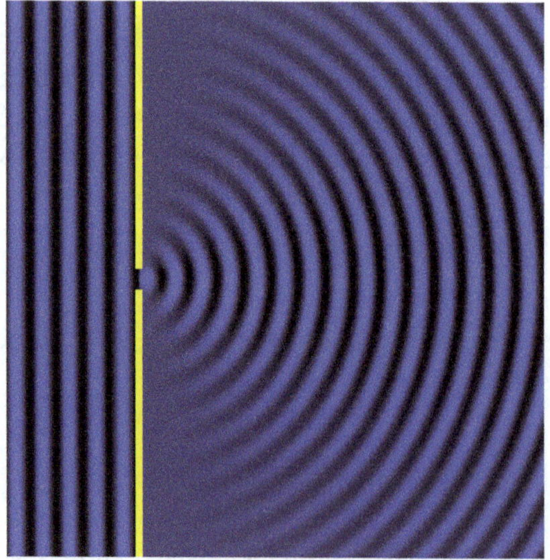

Waves propagating through a slit one wavelength wide

Diffraction processes affecting waves are amenable to quantitative description and analysis. Such treatments are applied to a wave passing through one or more slits whose width is specified as a proportion of the wavelength. Numerical approximations may be used, including the Fresnel and Fraunhofer approximations.

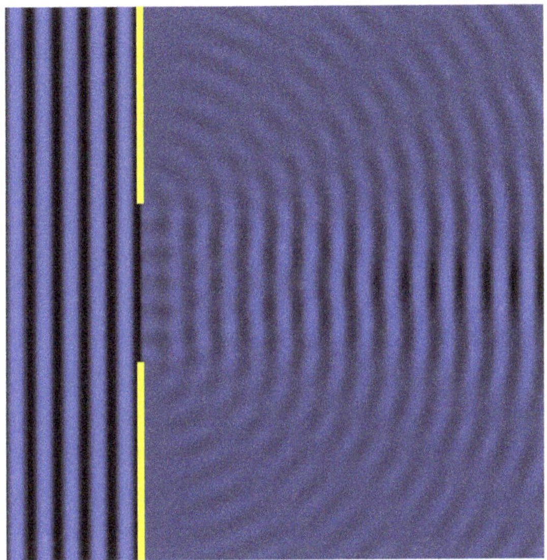

Waves propagating through a slit six wavelengths wide

General Diffraction

Because diffraction is the result of addition of all waves (of given wavelength) along all unobstructed paths, the usual procedure is to consider the contribution of an infinitesimally small neighborhood around a certain path (this contribution is usually called a wavelet) and then integrate over all paths (= add all wavelets) from the source to the detector (or given point on a screen).

Thus in order to determine the pattern produced by diffraction, the phase and the amplitude of each of the wavelets is calculated. That is, at each point in space we must determine the distance to each of the simple sources on the incoming wavefront. If the distance to each of the simple sources differs by an integer number of wavelengths, all the wavelets will be in phase, resulting in constructive interference. If the distance to each source is an integer plus one half of a wavelength, there will be complete destructive interference. Usually, it is sufficient to determine these minima and maxima to explain the observed diffraction effects.

The simplest descriptions of diffraction are those in which the situation can be reduced to a two-dimensional problem. For water waves, this is already the case, as water waves propagate only on the surface of the water. For light, we can often neglect one dimension if the diffracting object extends in that direction over a distance far greater than the wavelength. In the case of light shining through small circular holes we will have to take into account the full three-dimensional nature of the problem.

Several qualitative observations can be made of diffraction in general:

- The angular spacing of the features in the diffraction pattern is inversely proportional to the dimensions of the object causing the diffraction. In other words: the smaller the diffracting object, the wider the resulting diffraction pattern, and vice versa. (More precisely, this is true of the sines of the angles.)

- The diffraction angles are invariant under scaling; that is, they depend only on the ratio of the wavelength to the size of the diffracting object.
- When the diffracting object has a periodic structure, for example in a diffraction grating, the features generally become sharper. The fourth figure, for example, shows a comparison of a double-slit pattern with a pattern formed by five slits, both sets of slits having the same spacing between the center of one slit and the next.

Approximations

The problem of calculating what a diffracted wave looks like, is the problem of determining the phase of each of the simple sources on the incoming wave front. It is mathematically easier to consider the case of far-field or Fraunhofer diffraction, where the point of observation is far from that of the diffracting obstruction, and as a result, involves less complex mathematics than the more general case of near-field or Fresnel diffraction. To make this statement more quantitative, let us consider a diffracting object at the origin that has a size a. For definiteness let us say we are diffracting light and we are interested in what the intensity looks like on a screen a distance L away from the object. At some point on the screen the path length to one side of the object is given by the Pythagorean theorem

$$S=\sqrt{L^2+(x+a/2)^2}$$

If we now consider the situation where $L>>(x+a/2)$, the path length becomes

$$S\approx\left(L+\frac{(x+a/2)^2}{2L}\right)=L+\frac{x^2}{2L}+\frac{xa}{2L}+\frac{a^2}{8L}$$

This is the Fresnel approximation. To further simplify things: If the diffracting object is much smaller than the distance L, the last term will contribute much less than a wavelength to the path length, and will then not change the phase appreciably.

That is $\frac{a^2}{L}<<\lambda$. The result is the Fraunhofer approximation, which is only valid very far away from the object

$$S\approx L+\frac{x^2}{2L}+\frac{xa}{2L}$$

Depending on the size of the diffraction object, the distance to the object and the wavelength of the wave, the Fresnel approximation, the Fraunhofer approximation or neither approximation may be valid. As the distance between the measured point of diffraction and the obstruction point increases, the diffraction patterns or results predicted converge towards those of Fraunhofer diffraction, which is more often observed in nature due to the extremely small wavelength of visible light.

Diffraction from an Array of Narrow Slits

A Simple Quantitative Description

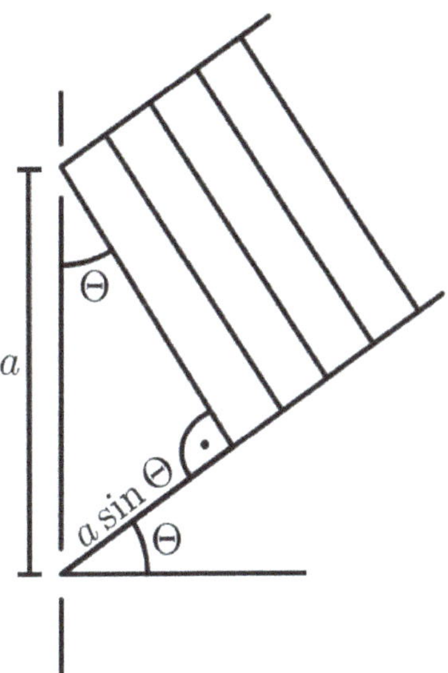

Diagram of a two slit diffraction problem, showing the angle to the first minimum, where a path length difference of a half wavelength causes destructive interference

Multiple-slit arrangements can be mathematically considered as multiple simple wave sources, if the slits are narrow enough. For light, a slit is an opening that is infinitely extended in one dimension, and this has the effect of reducing a wave problem in 3D-space to a simpler problem in 2D-space. The simplest case is that of two narrow slits, spaced a distance a apart. To determine the maxima and minima in the amplitude we must determine the path difference to the first slit and to the second one. In the Fraunhofer approximation, with the observer far away from the slits, the difference in path length to the two slits can be seen from the image to be

$$\Delta S = a \sin\theta$$

Maxima in the intensity occur if this path length difference is an integer number of wavelengths.

$$a \sin\theta = n\lambda$$

where

n is an integer that labels the *order* of each maximum,

λ is the wavelength,

a is the distance between the slits

and θ is the angle at which constructive interference occurs.

The corresponding minima are at path differences of an integer number plus one half of the wavelength:

$$a \sin\theta = \lambda(n + 1/2).$$

For an array of slits, positions of the minima and maxima are not changed, the *fringes* visible on a screen however do become sharper, as can be seen in the image.

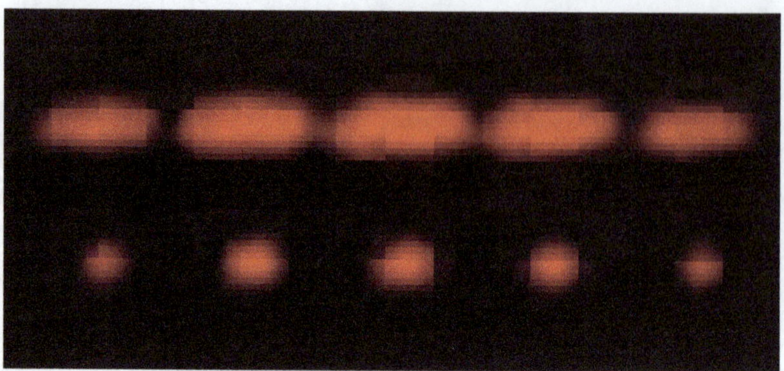

2-slit and 5-slit diffraction of red laser light

Mathematical Description

To calculate this intensity pattern, one needs to introduce some more sophisticated methods. The mathematical representation of a radial wave is given by

$$E(r) = A\cos(kr - \omega t + \phi) / r$$

where $k = \frac{2\pi}{\lambda}$, λ is the wavelength, ω is frequency of the wave and ϕ is the phase of the wave at the slits at time t=0. The wave at a screen some distance away from the plane of the slits is given by the sum of the waves emanating from each of the slits. To make this problem a little easier, we introduce the complex wave Ψ, the real part of which is equal to E

$$\Psi(r) = Ae^{i(kr - \omega t + \phi)} / r$$

$$E(r) = Re(\Psi(r))$$

The absolute value of this function gives the wave amplitude, and the complex phase of the function corresponds to the phase of the wave. Ψ is referred to as the complex amplitude. With N slits, the total wave at point x on the screen is

$$\Psi_{total} = Ae^{i(-\omega t + \phi)} \sum_{n=0}^{N-1} \frac{e^{ik\sqrt{(x-na)^2 + L^2}}}{\sqrt{(x-na)^2 + L^2}}.$$

Since we are for the moment only interested in the amplitude and relative phase, we can ignore any overall phase factors that are not dependent on x or n. We approximate $\sqrt{(x-na)^2 + L^2} \approx L + (x-na)^2 / 2L$. In the Fraunhofer limit we can neglect terms of order : $\frac{a^2}{2L}$ in the exponential, and any terms involving a / L or x / L in the denominator. The sum becomes

$$\Psi = A\frac{e^{i\left(k(\frac{x^2}{2L} + L) - \omega t + \phi\right)}}{L} \sum_{n=0}^{N-1} e^{-ik\frac{xna}{L}}$$

The sum has the form of a geometric sum and the can be evaluated to give

$$\Psi = A\frac{e^{i\left(k(\frac{x^2-(N-1)ax}{2L}+L)-\omega t+\phi\right)}}{L}\frac{\sin\left(\frac{Nkax}{2L}\right)}{\sin\left(\frac{kax}{2L}\right)}$$

The intensity is given by the absolute value of the complex amplitude squared

$$I(x) = \Psi\Psi^* = |\Psi|^2 = I_0\left(\frac{\sin\left(\frac{Nkax}{2L}\right)}{\sin\left(\frac{kax}{2L}\right)}\right)^2$$

where Ψ^* denotes the complex conjugate of Ψ.

Quantitative Analysis of Single-slit Diffraction

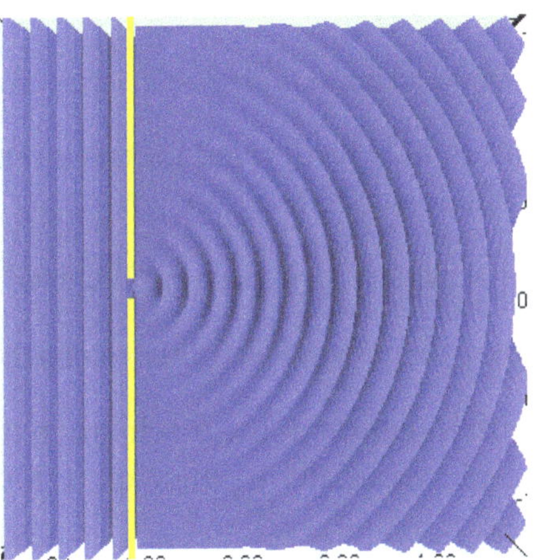

Numerical approximation of diffraction pattern from a slit of width equal to wavelength of an incident plane wave in 3D blue visualization

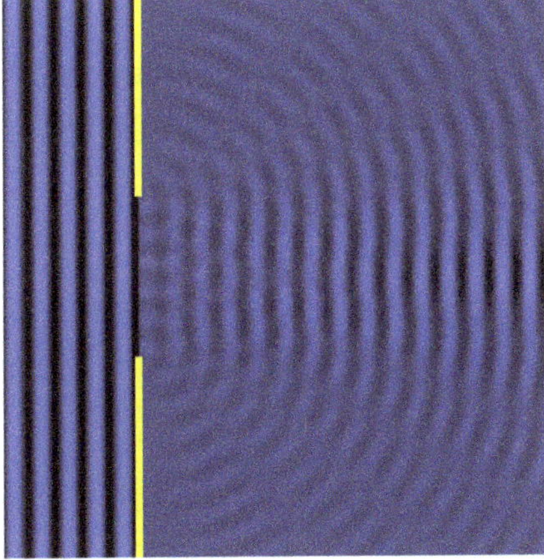

Numerical approximation of diffraction pattern from a slit of width equal to six time the wavelength of an incident plane wave in 3D blue visualization

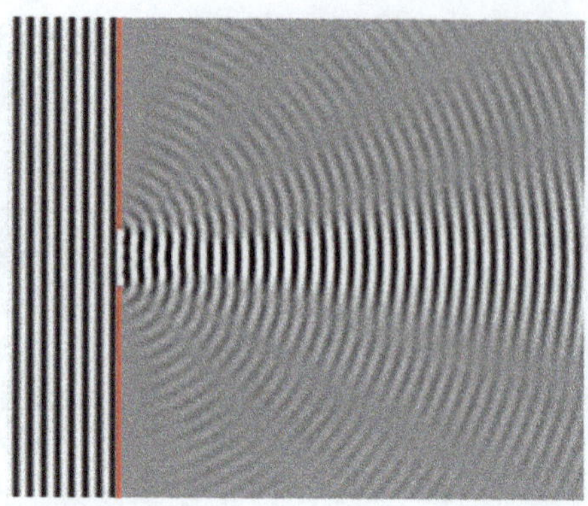

Numerical approximation of diffraction pattern from a slit of width four wavelengths with an incident plane wave. The main central beam, nulls, and phase reversals are apparent

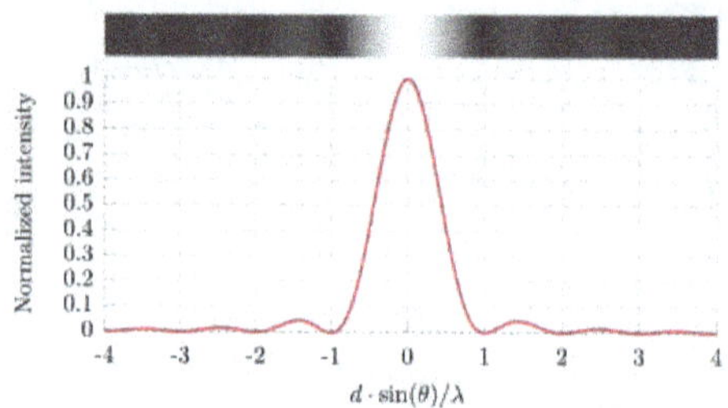

Graph and image of single-slit diffraction

As an example, an exact equation can now be derived for the intensity of the diffraction pattern as a function of angle in the case of single-slit diffraction.

A mathematical representation of Huygens' principle can be used to start an equation.

Consider a monochromatic complex plane wave Ψ' of wavelength λ incident on a slit of width a.

If the slit lies in the x′-y′ plane, with its center at the origin, then it can be assumed that diffraction generates a complex wave ψ, traveling radially in the r direction away from the slit, and this is given by:

$$\Psi = \int_{\text{slit}} \frac{i}{r\lambda} \Psi' e^{-ikr} \, d\text{slit}$$

Let (x′,y′,0) be a point inside the slit over which it is being integrated. If (x,0,z) is the location at which the intensity of the diffraction pattern is being computed, the slit extends from $x' = -a/2$ to $+a/2$, and from $y' = -\infty$ to ∞.

The distance r from the slot is:

$$r = \sqrt{\left(x - x'\right)^2 + y'^2 + z^2}$$

$$r = z\left(1 + \frac{\left(x - x'\right)^2 + y'^2}{z^2}\right)^{\frac{1}{2}}$$

Assuming Fraunhofer diffraction will result in the conclusion $z \gg |(x - x')|$. In other words, the distance to the target is much larger than the diffraction width on the target. By the binomial expansion rule, ignoring terms quadratic and higher, the quantity on the right can be estimated to be:

$$r \approx z\left(1 + \frac{1}{2}\frac{\left(x - x'\right)^2 + y'^2}{z^2}\right)$$

$$r \approx z + \frac{\left(x - x'\right)^2 + y'^2}{2z}$$

It can be seen that $1/r$ in front of the equation is non-oscillatory, i.e. its contribution to the magnitude of the intensity is small compared to our exponential factors. Therefore, we will lose little accuracy by approximating it as $1/z$.

$$\Psi = \frac{i\Psi'}{z\lambda}\int_{-\frac{a}{2}}^{\frac{a}{2}}\int_{-\infty}^{\infty} e^{-ik\left[z + \frac{\left(x - x'\right)^2 + y'^2}{2z}\right]} dy'\,dx'$$

$$= \frac{i\Psi'}{z\lambda} e^{-ikz} \int_{-\frac{a}{2}}^{\frac{a}{2}} e^{-ik\left[\frac{\left(x - x'\right)^2}{2z}\right]} dx' \int_{-\infty}^{\infty} e^{-ik\left[\frac{y'^2}{2z}\right]} dy'$$

$$= \Psi'\sqrt{\frac{i}{z\lambda}}\, e^{\frac{-ikx^2}{2z}} \int_{-\frac{a}{2}}^{\frac{a}{2}} e^{\frac{ikxx'}{z}} e^{\frac{-ikx'^2}{2z}} dx'$$

To make things cleaner, a placeholder 'C' is used to denote constants in the equation. It is important to keep in mind that C can contain imaginary numbers, thus the wave function will be complex. However, at the end, the ψ will be bracketed, which will eliminate any imaginary components.

Now, in Fraunhofer diffraction, kx'^2 / z is small, so $e^{\frac{-ikx'^2}{2z}} \approx 1$ (note that x' participates in this exponential and it is being integrated).

In contrast the term $e^{\frac{-ikx^2}{2z}}$ can be eliminated from the equation, since when bracketed it gives 1.

$$\langle e^{\frac{-ikx^2}{2z}} | e^{\frac{-ikx^2}{2z}} \rangle = e^{\frac{-ikx^2}{2z}} (e^{\frac{-ikx^2}{2z}})^* = e^{\frac{-ikx^2}{2z}} e^{\frac{+ikx^2}{2z}} = e^0 = 1$$

(For the same reason we have also eliminated the term e^{-ikz})

Taking $C = \Psi'\sqrt{\frac{i}{z\lambda}}$ results in:

$$\Psi = C\int_{-\frac{a}{2}}^{\frac{a}{2}} e^{\frac{ikxx'}{z}} dx'$$

$$= C\frac{\left(e^{\frac{ikax}{2z}} - e^{\frac{-ikax}{2z}}\right)}{\frac{ikx}{z}}$$

It can be noted through Euler's formula and its derivatives that $\sin x = \frac{e^{ix} - e^{-ix}}{2i}$ and $\sin\theta = \frac{x}{z}$.

$$\Psi = aC\frac{\sin\frac{ka\sin\theta}{2}}{\frac{ka\sin\theta}{2}} = aC\left[\operatorname{sinc}\left(\frac{ka\sin\theta}{2}\right)\right]$$

where the (unnormalized) sinc function is defined by $\operatorname{sinc}(x) \overset{\text{def}}{=} \frac{\sin(x)}{x}$.

Now, substituting in $\frac{2\pi}{\lambda} = k$, the intensity (squared amplitude) I of the diffracted waves at an angle θ is given by:

$$I(\theta) = I_0\left[\operatorname{sinc}\left(\frac{\pi a}{\lambda}\sin\theta\right)\right]^2$$

Quantitative Analysis of *N*-slit Diffraction

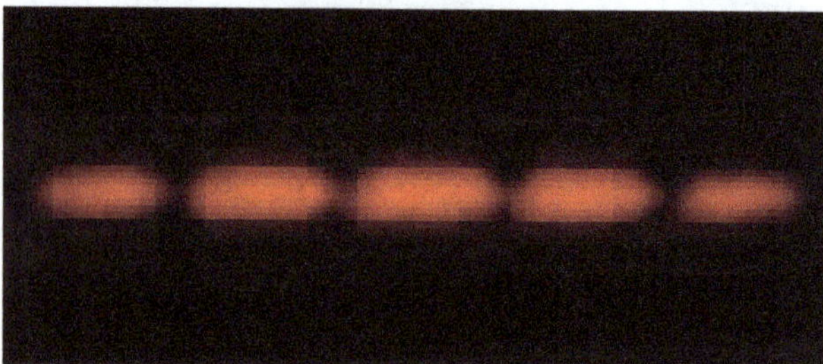

Double-slit diffraction of red laser light

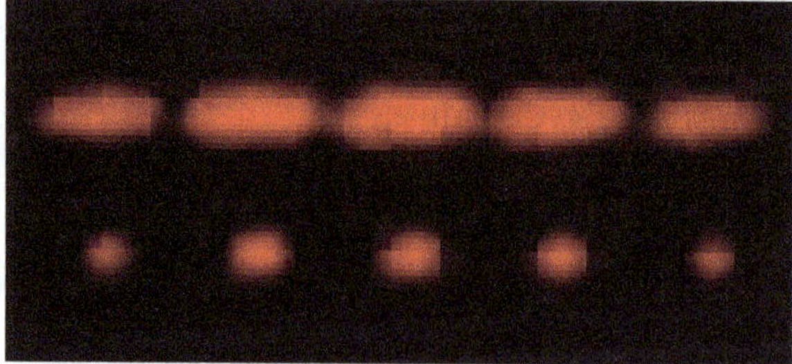

2-slit and 5-slit diffraction

Let us again start with the mathematical representation of Huygens' principle.

$$\Psi = \int_{\text{slit}} \frac{i}{r\lambda}\Psi' e^{-ikr}\, d\text{slit}$$

Consider *N* slits in the prime plane of equal size *a* and spacing *d* spread along the x′ axis. As above, the distance *r* from slit 1 is:

$$r = z\left(1+\frac{\left(x-x'\right)^2+y'^2}{z^2}\right)^{\frac{1}{2}}$$

To generalize this to N slits, we make the observation that while z and y remain constant, x′ shifts by

$$x'_{j=0\cdots n-1} = x'_0 - jd$$

Thus

$$r_j = z\left(1+\frac{\left(x-x'-jd\right)^2+y'^2}{z^2}\right)^{\frac{1}{2}}$$

and the sum of all N contributions to the wave function is:

$$\Psi = \sum_{j=0}^{N-1} C\int_{-\frac{a}{2}}^{\frac{a}{2}} e^{\frac{ikx\left(x'-jd\right)}{z}} e^{\frac{-ik\left(x'-jd\right)^2}{2z}} dx'$$

Again noting that $\frac{k\left(x'-jd\right)^2}{z}$ is small, so $e^{\frac{-ik(x'-jd)^2}{2z}} \approx 1$, we have:

$$\Psi = C\sum_{j=0}^{N-1}\int_{-\frac{a}{2}}^{\frac{a}{2}} e^{\frac{ikx\left(x'-jd\right)}{z}} dx'$$

$$= aC\sum_{j=0}^{N-1}\frac{\left(e^{\frac{ikax}{2z}-\frac{ijkxd}{z}} - e^{\frac{-ikax}{2z}-\frac{ijkxd}{z}}\right)}{\frac{2ikax}{2z}}$$

$$= aC\sum_{j=0}^{N-1} e^{\frac{ijkxd}{z}}\frac{\left(e^{\frac{ikax}{2z}} - e^{\frac{-ikax}{2z}}\right)}{\frac{2ikax}{2z}}$$

$$= aC\frac{\sin\frac{ka\sin\theta}{2}}{\frac{ka\sin\theta}{2}}\sum_{j=0}^{N-1} e^{ijkd\sin\theta}$$

Now, we can use the following identity

$$\sum_{j=0}^{N-1} e^{xj} = \frac{1-e^{Nx}}{1-e^{x}}.$$

Substituting into our equation, we find:

$$\Psi = aC\frac{\sin\frac{ka\sin\theta}{2}}{\frac{ka\sin\theta}{2}}\left(\frac{1-e^{iNkd\sin\theta}}{1-e^{ikd\sin\theta}}\right)$$

$$= aC\frac{\sin\frac{ka\sin\theta}{2}}{\frac{ka\sin\theta}{2}}\left(\frac{e^{-iNkd\frac{\sin\theta}{2}}-e^{iNkd\frac{\sin\theta}{2}}}{e^{-ikd\frac{\sin\theta}{2}}-e^{ikd\frac{\sin\theta}{2}}}\right)\left(\frac{e^{iNkd\frac{\sin\theta}{2}}}{e^{ikd\frac{\sin\theta}{2}}}\right)$$

$$= aC\frac{\sin\frac{ka\sin\theta}{2}}{\frac{ka\sin\theta}{2}}\frac{\frac{e^{-iNkd\frac{\sin\theta}{2}}-e^{iNkd\frac{\sin\theta}{2}}}{2i}}{\frac{e^{-ikd\frac{\sin\theta}{2}}-e^{ikd\frac{\sin\theta}{2}}}{2i}}\left(e^{i(N-1)kd\frac{\sin\theta}{2}}\right)$$

$$= aC\frac{\sin\left(\frac{ka\sin\theta}{2}\right)}{\frac{ka\sin\theta}{2}}\frac{\sin\left(\frac{Nkd\sin\theta}{2}\right)}{\sin\left(\frac{kd\sin\theta}{2}\right)}e^{i(N-1)kd\frac{\sin\theta}{2}}$$

We now make our k substitution as before and represent all non-oscillating constants by the I_0 variable as in the 1-slit diffraction and bracket the result. Remember that

$$\langle e^{ix} | e^{ix}\rangle = e^{0} = 1$$

This allows us to discard the tailing exponent and we have our answer:

$$I(\theta) = I_0\left[\operatorname{sinc}\left(\frac{\pi a}{\lambda}\sin\theta\right)\right]^2\cdot\left[\frac{\sin\left(\frac{N\pi d}{\lambda}\sin\theta\right)}{\sin\left(\frac{\pi d}{\lambda}\sin\theta\right)}\right]^2$$

General Case for far Field

In the far field, where r is essentially constant, then the equation:

$$\Psi = \int_{\text{slit}} \frac{i}{r\lambda}\Psi' e^{-ikr}\,d\text{slit}$$

is equivalent to doing a fourier transform on the gaps in the barrier.

Angular Resolution

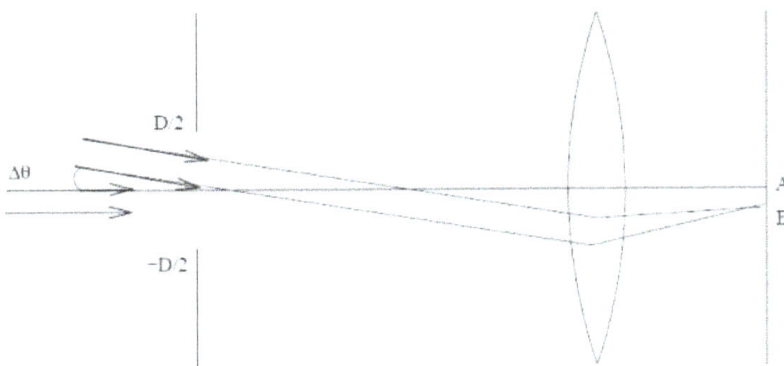

Light from two distant sources incident on single slit and their images

We consider a situation where the light from two distant sources is incident on a slit of dimension D. The sources are at an angular separation $\Delta\theta$ as shown in Figure. The light from the two sources is focused onto a screen. In the absence of diffraction the image of each source would be distinct point on the screen.

In reality we shall get the superposition of the diffraction patterns produced by the two sources as shown in Figure. The two diffraction patterns are centered on the positions A and B respectively where we expect the geometrical image. In case the angular separation $\Delta\theta$ is very small the two diffraction patterns will have a significant overlap. In such a situation it will not possible to make out that there are two sources as it will appear that there is a single source. Two sources at such small angular separations are said to be unresolved. The two sources are said to be resolved if their diffraction patterns do not have a significant overlap and it is possible to make out that there are two sources and not one. What is the smallest angle $\Delta\theta$ for which it is possible to make out that there are two sources and not one?

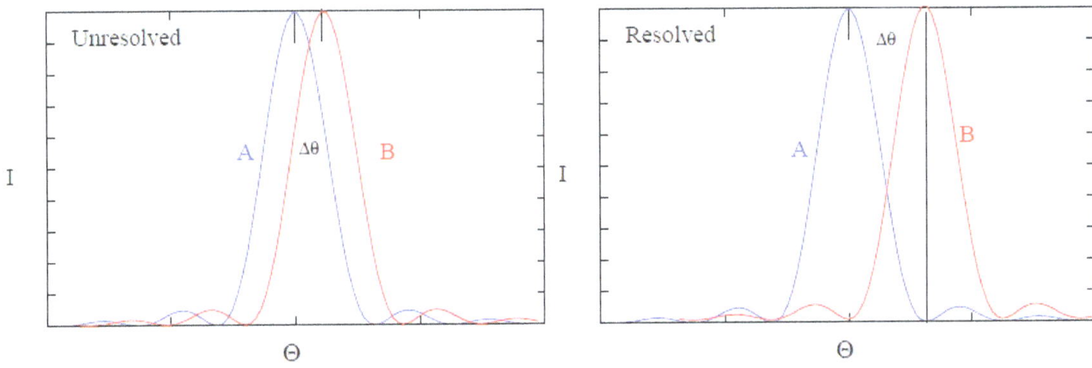

The Rayleigh criterion for resolution

Lord Rayleigh had proposed a criterion that the smallest separation at which it is possible to distinguish two diffraction patterns is when the maximum of one coincides with the first minimum of the other. It follows that two sources are resolved if their angular separation satisfies

$$\Delta\theta \geq \frac{\lambda}{D}$$

The smallest angular separation $\Delta\theta$ at which two sources are resolved is referred to as the "angular resolution" of the aperture. A slit of dimension D has an angular resolution of λ / D.

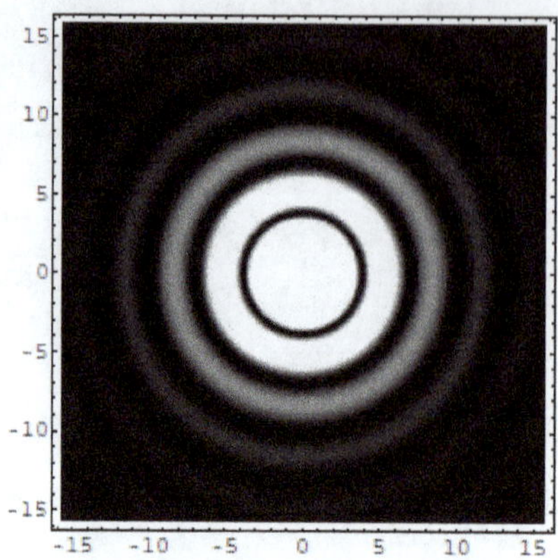

Circular aperture diffraction pattern

A circular aperture produces a circular diffraction pattern as shown in the Figure. The mathematical form is a little more complicated than the sinc function which appears when we have a rectangular aperture, but it is qualitatively similar. The first minima is at an angle $\theta = 1.22\lambda / D$ where D is the diameter of the aperture. It then follows that the "angular resolution" of a circular aperture is $1.22\lambda / D$. When a telescope of diameter D is used to observe a star, the image of the star is basically the diffraction pattern corresponding to the circular aperture of the telescope. Suppose there are two stars very close in the sky, what is the minimum angular separation at which it will be possible to distinguish the two stars? It is clear from our earlier discussion that the two stars should be at least $1.22\lambda / D$ apart in angle, other they will not be resolved. The Figure shows not resolved, barely resolved and well resolved cases for a circular aperture.

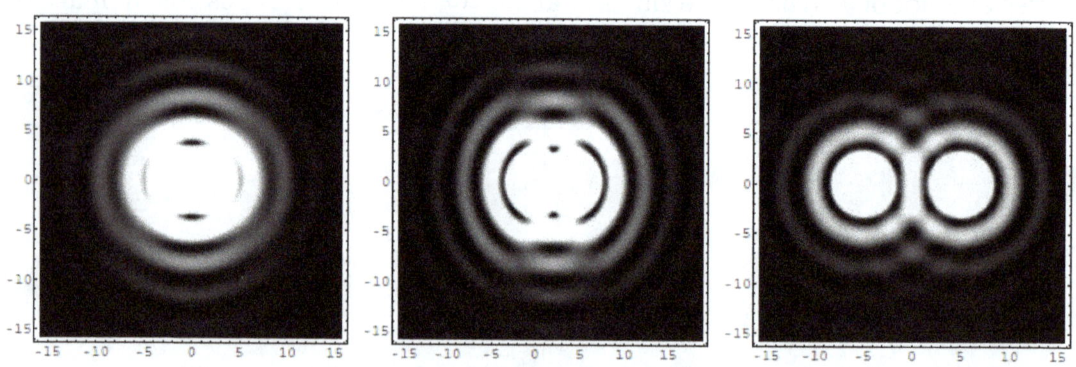

Not resolved, barely resolved and well resolved cases

Diffraction determines the angular resolution of any imaging instrument. This is typically of the order of λ / D where D is the size of the instrument's aperture.

Double Slit Diffraction Pattern

In this section we consider the Fraunhofer diffraction pattern for two nearby slits. The difference between the double slit diffraction and the Young's two slit interference is the individual slit width. In the former case the slits are thin but not of negligible widths but in the latter the two slits are ideal and infinitely thin or of negligible widths.

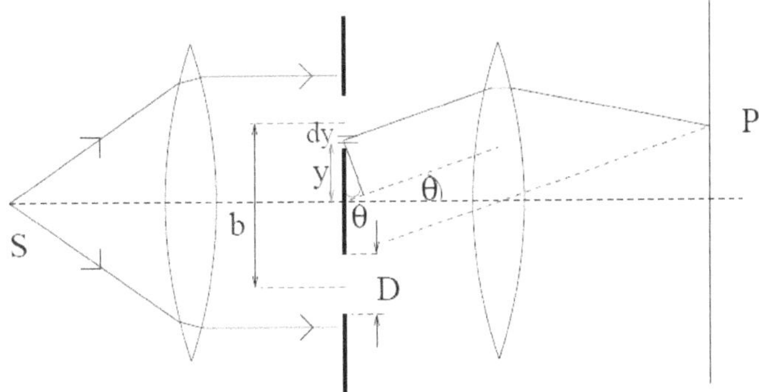

Double slit diffraction

We refer to the Figure and proceed like the single slit case to obtain the intensity pattern by summing the amplitudes. In this case we take individual slit widths as D and the separation between the two slits as b. Summing the amplitudes we have

$$\tilde{E} = \int dA = \tilde{A}\left(\int_{-b/2-D/2}^{-b/2+D/2} + \int_{b/2-D/2}^{b/2+D/2}\right)\exp\left(i2\pi y \sin\theta / \lambda\right)dy,$$

and the corresponding intensity distribution is given by

$$I = \tilde{E}\tilde{E}^* = I_0 \cos^2 \gamma \frac{\sin^2 \beta}{\beta^2}.$$

Where $\beta = \pi D \sin\theta / \lambda$ *and* $\gamma = \pi b \sin\theta / \lambda$.

It can be easily seen that in the limit $b \to 0$ the result co-insides with the single slit result whereas the limit $D \to 0$ recovers the Young's two-slit interference result discussed in earlier chapters. A typical diffraction pattern for a double slit is shown in Figure. The intensity pattern of a double slit is the product of the diffraction pattern of a single slit and the interference pattern of an ideal Young's double slit as shown in the Figure.

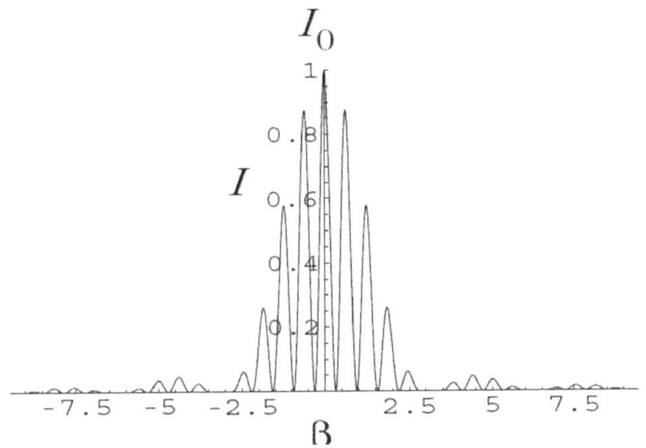

A double slit diffraction pattern

If one considers only the single slit diffraction, this gives minimum intensity when the numerator of the last factor becomes zero. This always is the case when the following applies,

$$D \sin\theta = m\lambda, \qquad m = 1,2,3,\cdots$$

Principal maximum is obtained for $\theta = 0$, when the last factor of the expression becomes 1. There are other secondary maxima approximately at $\theta = 1.43°, 2.46°, 3.47°, 4.48°$ etc. On the other hand, the double slit interference pattern has maxima at angles satisfying the condition,

$$b \sin\theta = n\lambda, \qquad n = 0,1,2,3,\cdots$$

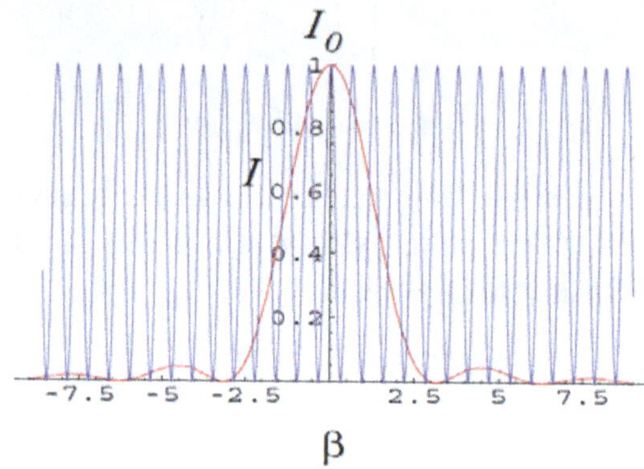

Single slit diffraction pattern on Young double slit interference

An interesting situation arises when both equations, viz. equations are satisfied for a particular value of θ. This happens when,

$$\frac{b}{D} = \frac{n}{m}$$

In this case order n^{th} interference maximum will have zero intensity and will therefore be missing. In the Figure, $b/D = 5$, hence we find that the 5th secondary peak is missing. Condition is again satisfied for $n = 10$ and $m = 2$, so the 10th peak is again absent. Similarly 15th, 20th etc. peaks will also be missing in this case. These are known as missing orders in the literature. Missing orders of double slit fall at multiples of b/D. The central envelope has 9 peaks within it, whereas the side lobes will have 4 peaks each for the case shown above.

Chain of Sources

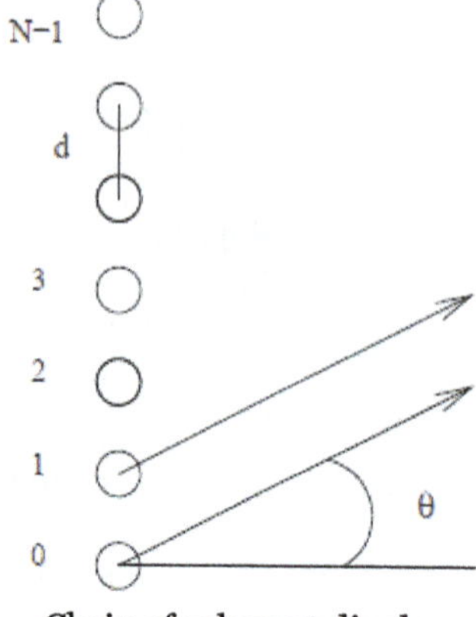

Chain of coherent dipoles

Consider N dipole oscillators arranged along a linear chain as shown in Figure, all emitting radiation with identical amplitude and phase. How much radiation will a distant observer at an angle θ receive? If $\tilde{E}_0, \tilde{E}_1, \tilde{E}_2, \ldots,$ and $\tilde{E}_{N-1}$ are the radiations from the 0th, 1st, 2nd, ...,and the N −1th oscillator respectively, $\tilde{E}_1$ is identical to $\tilde{E}_0$ except for a phase difference as it travels a shorter path. We have

$$\tilde{E}_1 = \tilde{E}_0 e^{i2\alpha}$$

where $2\alpha = \frac{2\pi}{\lambda} d \sin\theta$ is the phase difference that arises due to the path difference.

Similarly $\tilde{E}_2 = \left[e^{i2\alpha}\right]^2 \tilde{E}_0$. The total radiation is obtained by summing the contributions from all the sources and we have

$$\tilde{E} = \sum_{n=0}^{N-1} \tilde{E}_n = \sum_{n=0}^{N-1} \left[e^{i2\alpha}\right]^n \tilde{E}_0$$

This is a geometric progression, on summing this we have

$$\tilde{E} = \tilde{E}_0 \frac{e^{i2N\alpha} - 1}{e^{i2\alpha} - 1}.$$

This can be simplified further

$$\tilde{E} = \tilde{E}_0 \frac{e^{iN\alpha}}{e^{i\alpha}} \frac{e^{iN\alpha} - e^{-iN\alpha}}{e^{i\alpha} - e^{-i\alpha}} = \tilde{E}_0 e^{i(N-1)\alpha} \frac{\sin(N\alpha)}{\sin(\alpha)}$$

which gives the intensity to be

$$I = 0.5\tilde{E}\tilde{E}^* = I_0 \frac{\sin^2(N\alpha)}{\sin^2(\alpha)}$$

Where

$$\alpha = \frac{\pi d \sin\theta}{\lambda}$$

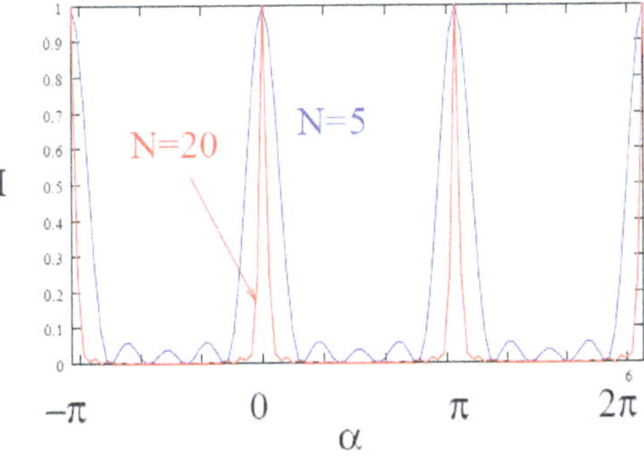

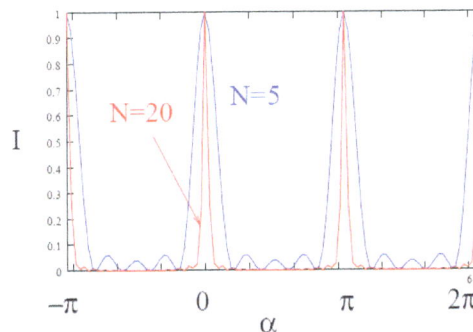

Intensity pattern for chain of dipoles

Plotting the intensity as a function of α we see that it has a value $I = N^2 I_0$ $at\ \alpha = 0$. Further, it has the same value $I = I_0 N^2$ at all other α values where both the numerator and denominator are zero i.e $\alpha = m\pi\,(m = 0, \pm 1, \pm 2, \ldots)$ or

$$d\sin\theta = m\lambda\,(m = 0, \pm 1, \pm 2, \pm 3 \ldots)$$

The intensity is maximum whenever this condition is satisfied. These are referred to as the primary maxima of the diffraction pattern and m gives the order of the maximum.

The intensity drops away from the primary maxima. The intensity becomes zero $N-1$ times between any two successive primary maxima and there are $N-2$ secondary maxima in between. The number of secondary maxima increases and the primary maxima becomes increasingly sharper if the number of sources N is increased. Let us estimate the width of the mth order principal maximum. The mth order principal maximum occurs at an angle θ_m which satisfies,

$$d\sin\theta_m = m\lambda.$$

If $\Delta\theta_m$ is the width of the maximum, the intensity should be zero at $\theta_m + \Delta\theta_m$ ie.

$$\frac{\pi d \sin(\theta_m + \Delta\theta_m)}{\lambda} = m\pi + \frac{\pi}{N}$$

which implies that

$$\sin(\theta m + \Delta\theta_m) = \frac{\lambda}{d}\left(m + \frac{1}{N}\right)$$

Expanding

$$\sin(\theta_m + \Delta\theta_m) = \sin\theta_m \cos\Delta\theta_m + \cos\theta_m \sin\Delta\theta_m$$

and assuming that $\Delta\theta_m \ll 1$ we have

$$\Delta\theta_m \cos\theta_m = \frac{\lambda}{dN}$$

which gives the width to be

$$\Delta\theta_m = \frac{\lambda}{dN\cos\theta_m}.$$

Thus we see that the principal maxima get sharper as the number of sources increases. Further, the 0th order maximum is the sharpest, and the width ofthe maximum increases with increasing order m.

The chain of radiation sources serves as an useful model for many applications.

Phased Array

Consider first any one of the dipole radiators shown in Figure. For a dipole oriented perpendicular to the plane of the page, the radiation is uniform in all directions on the plane of the page. Suppose we want to construct something like a radar that sends out radiation in only a specific direction, and not in all directions. It is possible to do so using a chain of N dipoles with spacing $d < \lambda$. The maxima of the radiation pattern occur at

$$\sin\theta = m\frac{\lambda}{d}$$

and as $\lambda / d > 1$ the only solutions are at $\theta = 0$ *and* $\theta = \pi$ (i.e. for $m = 0$).Thus the radiation is sent our only in the forward and backward directions, the radiation from the different dipoles cancel out in all other directions. The width of this maxima is

$$\Delta\theta = \frac{\lambda}{Nd}$$

which is determined by the separation between the two extreme dipoles in the chain.

The direction at which the maximum radiation goes out can be changed by introducing a constant phase difference 2ϕ between every pair of adjacent oscillators. The phase difference in the radiation received from any two adjacent dipoles is now given by

$$2\alpha = \frac{2\pi}{\lambda} d \sin\theta + 2\phi$$

and the condition at which the maxima occurs is now given by

$$\sin\theta = \left(\frac{\lambda}{d}\right)\left[m + \frac{\phi}{\pi}.\right]$$

The device discussed here is called a “phased array”, and it can be used to send out or receive radiation from only a specific region of the sky. This has several applications in communications, radars and radio-astronomy.

Diffraction Grating

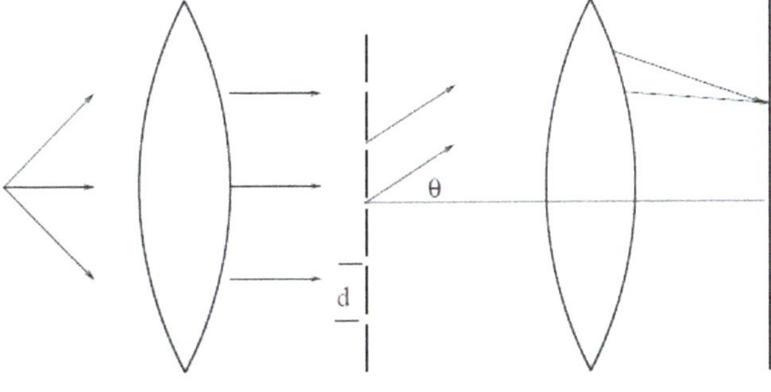

Diffraction grating

We consider the transmission diffraction grating shown in Figure. The transmission grating is essentially a periodic arrangement of N slits, each slit of width D and slit spacing d. The spacing between successive slits d is referred to as the "grating element" or as the "period of the grating". Each slit acts like a source, and the diffraction grating is equivalent to the chain of sources shown in Figure.

The intensity pattern of a diffraction grating is the product of the intensity pattern of a single slit and the intensity pattern of a periodic arrangement of emmiters

$$I = I_0 \frac{\sin^2(N\alpha)}{\sin^2(\alpha)} \sin c^2(\beta)$$

Where

$$\alpha = \frac{\pi d \sin\theta}{\lambda} \quad and \quad \beta = \frac{\pi D \sin\theta}{\lambda}$$

Typically the slit spacing d is larger than the slit width *i.e.* $d > D$. Figure shows the intensity pattern for a diffraction grating. The finite slit width

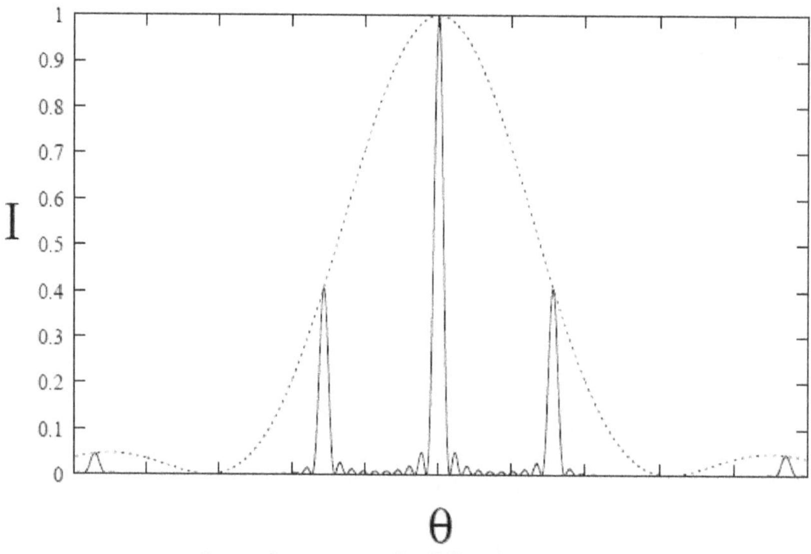

Intensity pattern of a diffraction grating

causes the higher order primary maximas to be considerably fainter than the low order ones.

The transmission grating is a very useful device in spectroscopy. The grating is very effective in dispersing the light into different wavelength components. For each wavelength the *mth* order primary maximum occurs at a different angle determined by

$$\sin\theta_m = m\frac{\lambda}{d}$$

The diffraction pattern, when two different wavelengths are incident on a grating, is shown in the Figure.

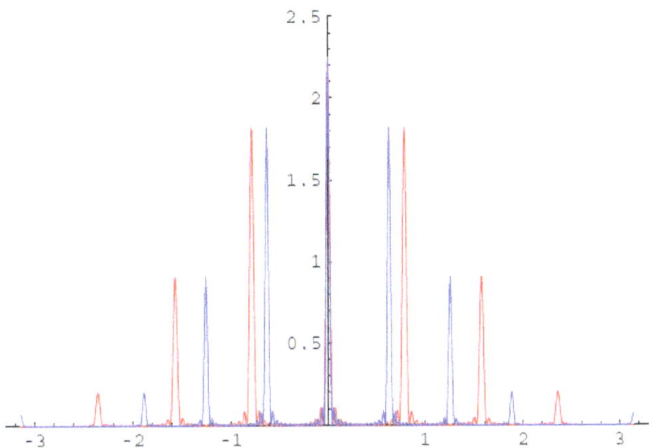

Diffraction pattern for two wavelengths (Intensity vs β)

The dispersive power of a grating is defined as

$$D = \left(\frac{d\theta_m}{d\lambda}\right) = \frac{m}{d\cos\theta_m}$$

We see that it increases with the order m and is inversely proportional to d. The finer the grating (small d) the more its dispersive power. Also, the higher orders have a greater dispersive power, but the intensity of these maxima is also fainter.

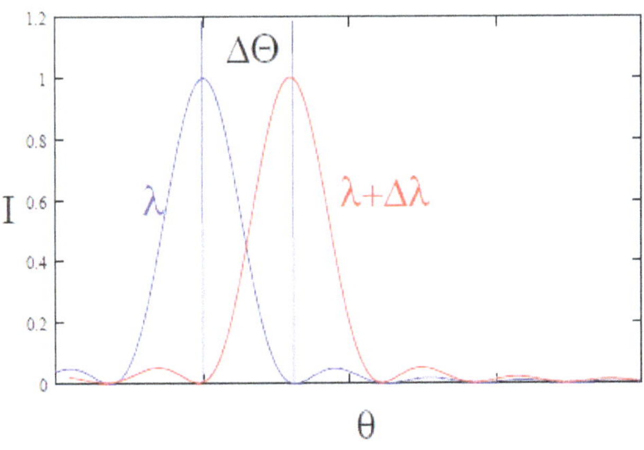

Chromatic resolution

The Chromatic Resolving Power (CRP) quantifies the ability of a grating to resolve two spectral lines of wavelengths λ *and* $\lambda + \Delta\lambda$. Applying Rayleigh's criterion, it will be possible to resolve the lines if the maximum of one coincides with the minimum of the other.

The first minimum corresponding to λ is seperated from the maximum of λ by an angle.

$$\Delta\theta = \frac{\lambda}{Nd\cos\theta_m},$$

The maxima corresponding to λ and $\lambda + \Delta\lambda$ are seperated by an angle.

$$\Delta\theta = \frac{m\Delta\lambda}{d\cos\theta_m}.$$

Equating these gives the CRP to be

$$R \equiv \frac{\lambda}{\Delta\lambda} = Nm$$

The chromatic resolving power increases with the number of surfs or rulings in the grating. It also increase with the order of spectrum. This makes the grating a very powerful dispersive element in spectrometers.

X-ray Diffraction

We have seen that a one-dimensional periodic arrangement of coherent radiation sources (chain of sources) produces a diffraction pattern. The diffraction grating is an example. Atoms and molecules have a three dimensional periodic arrangement inside crystalline solids. A diffraction pattern is produced if the atoms or molecules act like a three dimensional grating. Inside crystalline solids the inter-atomic spacing is of the order of 1 A. Crystals can produce a

diffraction pattern with X-ray whose wavelength is comparable to inter atomic spacings. The wavelength of visible light is a few thousand times larger and this does not serve the purpose.

X-ray is incident on a crystal as shown in Figure. The oscillating electric field of this electromagnetic wave induces a oscillating dipole moment in every atom or molecule inside the crystal. These dipoles oscillate at the same frequency as the incident X-ray. The oscillating dipoles emit radiation in all directions at the same frequency as the incident radiation, this is known as Thomson scattering. Every atom scatters the incident X-ray in all directions.

The radiation scattered from different atoms is coherent. The total radiation scattered in any particular direction is calculated by superposing the contribution from each atom.

For a crystal where the atoms have a periodic arrangement, it is convenient to think of the three-dimensional grating as a set of planes arranged in a onedimensional grating as shown in Figure.

Consider X-ray incident at a grazing angle of θ as shown in Figure. The intensity of the reflected X-ray will be maximum when the phase difference

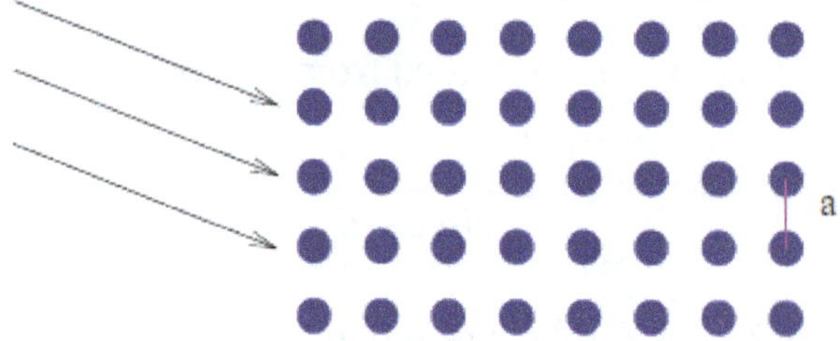

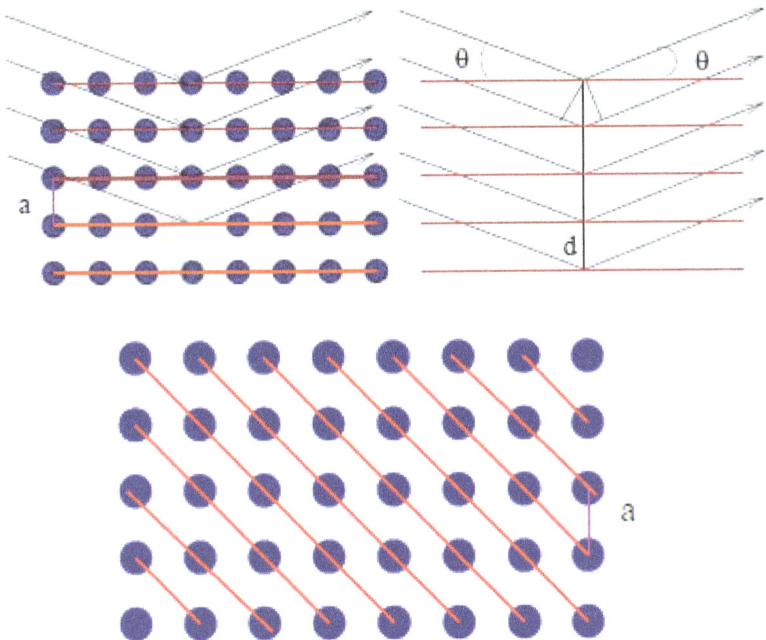

between the waves reflected from two successive planes is λ or its integer multiple. This occurs when

$$2d\sin\theta = m\lambda.$$

This formula is referred to as Bragg's Law. The diffraction can also occur from other planes as shown in Figure.

The spacing between the planes d is different in the two cases and the maxima will occur at a different angle. The first set of planes are denoted by the indices (1, 0, 0) and the second set by (1, 1, 0). It is, in principle, possible to have a large number of such planes denoted by the indices (h,k,l) referred to as the Miller indices. The distance between the planes is

$$d(h,k,l) = \frac{a}{\sqrt{h^2 + k^2 + l^2}}$$

where is the lattice constant or lattice spacing. Note that the crystal has been assumed to be cubic.

Figure shows a schematic diagram of an X-ray diffractometer. This essentially allows us to measure the diffracted X-ray intensity as a function of 2θ as shown in Figure.

Figure shows the unit cell of $La_{0.66}Sr_{0.33}MnO_3$. X-ray of wavelength λ. = 1.542A is used in an X-ray diffractometer, the resulting diffraction pattern with intensity as a function of 2θ is shown in Figure.

The 2θ values of the first three peaks have been tabulated below. The question is how to interpret the different peaks. All the peaks shown correspond to $m = 1$ *ie.* first order diffraction maximas, the higher orders $m = 2,3,\ldots$ are much fainter. The different peaks correspond to different Miller indices which give different values of d. The maxima at the smallest θ arises from the largest d value which correspons to the indices (h,k,l) = (1, 0, 0). The other maxima may be interpreted using the fact that θ and d are inversely related.

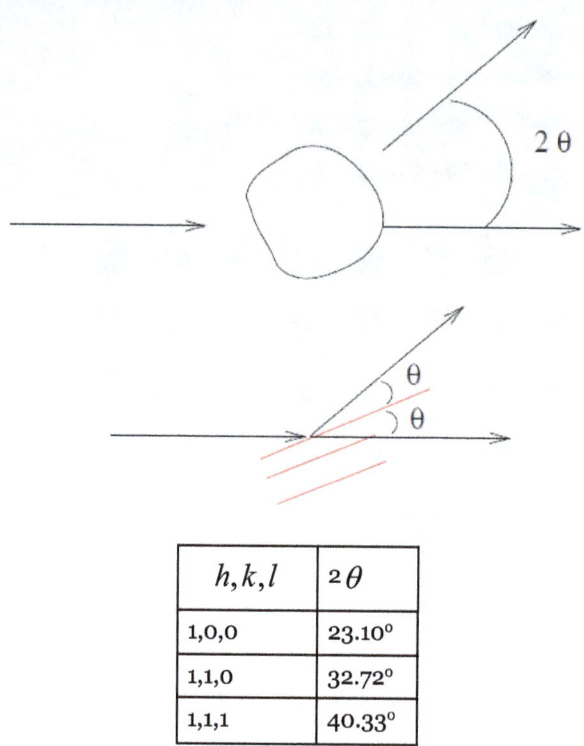

h,k,l	2θ
1,0,0	23.10°
1,1,0	32.72°
1,1,1	40.33°

X-ray Scattering Techniques

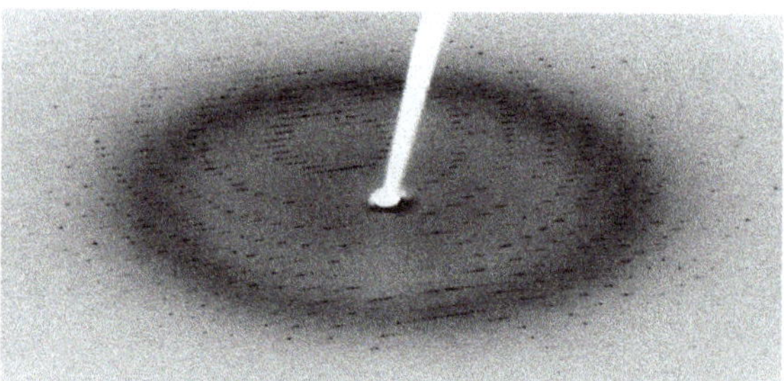

This is an X-ray diffraction pattern formed when X-rays are focused on a crystalline material, in this case a protein. Each dot, called a reflection, forms from the coherent interference of scattered X-rays passing through the crystal

X-ray scattering techniques are a family of non-destructive analytical techniques which reveal information about the crystal structure, chemical composition, and physical properties of materials and thin films. These techniques are based on observing the scattered intensity of an X-ray beam hitting a sample as a function of incident and scattered angle, polarization, and wavelength or energy.

Note that X-ray diffraction is now often considered a sub-set of X-ray scattering, where the scattering is elastic and the scattering object is crystalline, so that the resulting pattern contains sharp spots analyzed by X-ray crystallography (as in the Figure). However, both scattering and diffraction are related general phenomena and the distinction has not always existed. Thus Guinier's classic text from 1963 is titled "X-ray diffraction in Crystals, Imperfect Crystals and Amorphous Bodies" so 'diffraction' was clearly not restricted to crystals at that time.

Scattering Techniques

Elastic Scattering

- X-ray diffraction or more specifically Wide-angle X-ray diffraction (WAXD)
- Small-angle X-ray scattering (SAXS) probes structure in the nanometer to micrometer range by measuring scattering intensity at scattering angles 2θ close to 0°.
- X-ray reflectivity is an analytical technique for determining thickness, roughness, and density of single layer and multilayer thin films.
- Wide-angle X-ray scattering (WAXS), a technique concentrating on scattering angles 2θ larger than 5°.

Inelastic X-ray Scattering (IXS)

In IXS the energy and angle of inelastically scattered X-rays are monitored, giving the dynamic structure factor $S(\mathbf{q},\omega)$. From this many properties of materials can be obtained, the specific property depending on the scale of the energy transfer. The table below, listing techniques, is adapted from. Inelastically scattered X-rays have intermediate phases and so in principle are not useful for X-ray crystallography. In practice X-rays with small energy transfers are included with the diffraction spots due to elastic scattering, and X-rays with large energy transfers contribute to the background noise in the diffraction pattern.

Technique	Typical Incident Energy, keV	Energy transfer range, eV	Information on:
Compton scattering	100	1,000	Fermi Surface Shape
Resonant IXS (RIXS)	4-20	0.1 - 50	Electronic Structure & Excitations
Non-Resonant IXS (NRIXS)	10	0.1 - 10	Electronic Structure & Excitations
X-ray Raman scattering	10	50 - 1000	Absorption Edge Structure, Bonding, Valence
High resolution IXS	10	0.001 - 0.1	Atomic Dynamics,Phonon Dispersion

Beats

In this chapter we consider the superposition of two waves of different frequencies. At a fixed point along the propagation direction of the waves, the time evolution of the two wave are,

$$\tilde{A}_1(t) = A_0 e^{i\omega_1 t},$$

And

$$\tilde{A}_2(t) = A_0 e^{i\omega_2 t},$$

The superposition of these two waves of equal amplitude is

$$\tilde{A}(t) = \tilde{A}_1(t) + \tilde{A}_2(t),$$

which can be written as,

$$\tilde{A}(t) = A_0\left[e^{-i(\omega_2-\omega_1)t/2}e^{i(\omega_1+\omega_2)t/2} + e^{i(\omega_2-\omega_1)t/2}e^{i(\omega_1+\omega_2)t/2}\right],$$

$$= 2A_0 \cos(\Delta\omega t/2)e^{i\omega t},$$

where $\Delta\omega = \omega_2 - \omega_1$ *and* $\bar{\omega} = (\omega_1 + \omega_2)/2$ are respectively the difference and the average of the two frequencies. If the two frequencies ω_1 and ω_2 are very close, and the difference in frequencies $\Delta\omega$ is much smaller than $\bar{\omega}$, we can think of the resultant as a fast varying wave with frequency $\bar{\omega}$ whose amplitude varies slowly at a frequency $\Delta\omega/2$. The intensity of the resulting wave is modulated at a frequency $\Delta\omega$. This slow modulation of the intensity is referred to as beats. This modulation is heard when two strings of a musical instrument are nearly tuned and this is useful in tuning musical instruments.

In the situation where the two amplitudes are different we have,

$$\tilde{A}(t) = a_1e^{i\omega_1 t} + a_2e^{i\omega_2 t} = \left[a_1e^{i\Delta\omega t/2} + a_2e^{i\Delta\omega t/2}\right]e^{i\omega t}.$$

Again we see that we have a fast varying component whose amplitude is modulated slowly. The intensity of the resultant wave is,

$$I = AA^* = a_1^2 + a_2^2 + 2a_1a_2 \cos(\Delta\omega t).$$

We see that the intensity oscillates at the frequency difference $\Delta\omega$, and it never goes to zero if the two amplitudes are different.

Radio transmission based on "amplitude modulation" is the opposite of this. The transmitter has a generator which produces a sinusoidal electrical wave at a high frequency, say 800 kHz which is the transmission frequency. This is called the carrier wave. The signal which is to be transmitted, say sound, is a relatively slowly varying signal in the frequency range 20 Hz to 20 kHz. The sound is converted to an electrical signal and the amplitude of the carrier wave is modulated by the slowly varying signal. Mathematically,

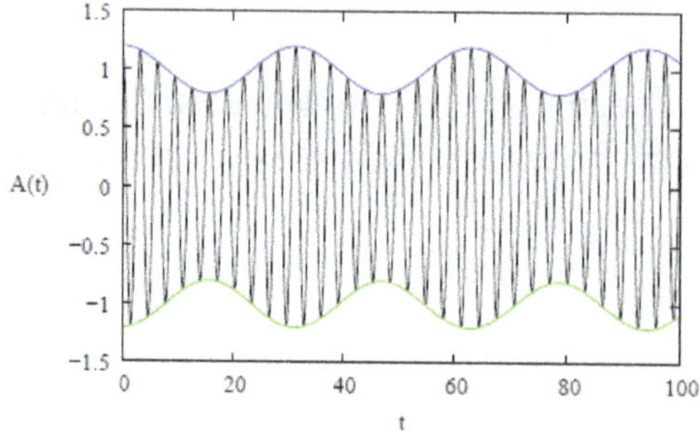

A modulated wave

$$\tilde{A}(t) = \left[1 + f(t)\right] e^{i\omega_c t},$$

where ω_c is the angular frequency of the carrier wave, and $f(t)$ is the slowly varying signal. As an example we consider a situation where the signal has a single frequency component,

$$f(t) = a_m \cos(\omega_m t),$$

where a_m is the amplitude and ω_m the frequency of the modulating signal. The transmitted signal $A(t)$ is shown in Figure. The envelope contains the signal. This can be recovered at the receiver by discarding the carrier and retaining only the envelope. The transmitted signal can be expressed as,

$$A(t) = e^{i\omega_c t} + \frac{a_m}{2} e^{i(\omega_c + \omega_m)t} + \frac{a_m}{2} e^{i(\omega_c - \omega_m)t}.$$

We see that though the transmitter originally produces output only at a single frequency ω_c when there is no modulation, it starts transmitting two other frequencies $\omega_c + \omega_m$ and $\omega_c - \omega_m$ once the signal is modulated. These new frequencies are referred to as sidebands. If we plot the spectrum of the radiation from the transmitter, we see that the energy is distributed in three frequencies as shown in Figure.

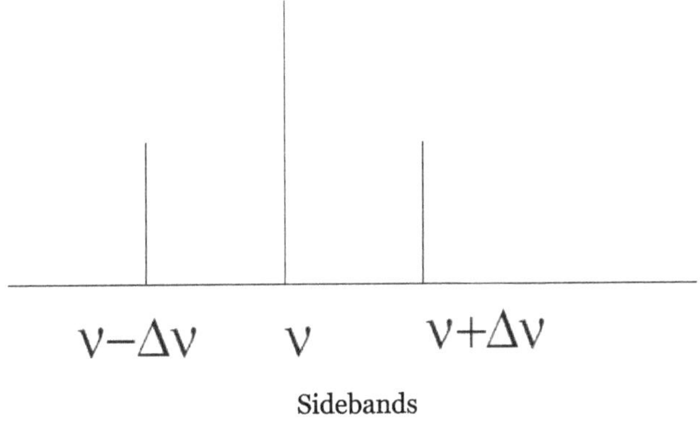

Sidebands

For a more complicated sound signal, the sidebands will be spread over a range of frequencies instead of a few discrete frequencies. The audible frequency range extends upto 20KHz but the transmitters and receivers usually do not work beyond 10KHz. So a radio station transmitting at 800KHz will actually be transmitting modulated signal in the frequency range 790KHz to 810KHz.

If our radio receiver were so sensitive that it picks up only a very small range of frequency around 800KHz, we would not be able to hear the sound that is being transmitted as the higher frequency components would be missing.

Further, if there were two stations one at 800KHz and another at 805KHz, the transmissions from the two stations would overlap and we would get a garbage sound from our receivers. The stations should transmit at frequencies that are sufficiently apart so that they do not overlap. Typically the frequency range 500KHz to 1500KHz is available for AM transmission and it is possible to accommodate a large number of stations.

We next consider the full position and time dependence of the superposition of two waves of different frequencies. Assuming equal amplitudes for the two waves we have,

$$\mathrm{A}(t) = A\left[e^{i(\omega_1 t - k_1 x)} + e^{i(\omega_2 t - k_2 x)}\right].$$

Proceeding in exactly the same way as when we considered only the time dependence, we now have,

$$\mathrm{A}(t) = 2A\cos\left(\frac{\Delta\omega}{2}t - \frac{\Delta k}{2}x\right)e^{i(\bar{\omega}t - \bar{k}x)},$$

where $\bar{\omega} = (\omega_2 + \omega_1)/2$ and $\bar{k} = (k_2 + k_1)/2$ are the mean angular frequency and wave number respectively, and $\Delta\omega = \omega_2 - \omega_1$ and $\Delta k = k_2 - k_1$ are the difference in the angular frequency and wave number respectively.

Let us consider a situation where the two frequencies are very close such that $\Delta\omega \ll \bar{\omega}$ *and* $\Delta k \ll \bar{k}$ the resultant can then be interpreted as a travelling wave with angular frequency and wave number $\bar{\omega}$ and $\bar{k}$ respectively. This wave has a phase velocity,

$$\upsilon_p = \bar{\omega}/\bar{k}.$$

The amplitude of this wave undergoes a slow modulation. The modulation itself is a travelling wave that propagates at a speed $\frac{\Delta\omega}{\Delta k}$. The speed at which the modulation propagates is called the group velocity, and we have

$$\upsilon_g = \frac{d\omega}{dk}.$$

As discussed earlier, it is possible to transmit signals using waves by modulating the amplitude. Usually (but not always) signals propagate at the group velocity.

There are situations where the phase velocity is greater than the speed of light in vacuum, but the group velocity usually comes out to be smaller. In all cases it is found that no signal propagates at a speed faster than the speed of light in vacuum. This is one of the fundamental assumptions in Einstein's Theory of Relativity.

References

- Brezger, B.; Hackermüller, L.; Uttenthaler, S.; Petschinka, J.; Arndt, M.; Zeilinger, A. (February 2002). "Matter–Wave Interferometer for Large Molecules". Physical Review Letters. 88 (10): 100404. Bibcode:2002PhRvL..88j0404B. PMID 11909334. arXiv:quant-ph/0202158. doi:10.1103/PhysRevLett.88.100404. Retrieved 2007-04-30
- Halliday, David; Resnick, Robert; Walker, Jerl (2005), Fundamental of Physics (7th ed.), USA: John Wiley and Sons, Inc., ISBN 0-471-23231-9
- Rashed, Roshdi (1990). "A pioneer in anaclastics: Ibn Sahl on burning mirrors and lenses". Isis. 81 (3): 464–491. doi:10.1086/355456

- Baron, A. Q. R. "Introduction to High-Resolution Inelastic X-Ray Scattering" (PDF). arxiv.org. Retrieved 27 September 2016
- Ayahiko Ichimiya; Philip I. Cohen (13 December 2004). Reflection High-Energy Electron Diffraction. Cambridge University Press. ISBN 978-0-521-45373-8
- Thomas Young (1804-01-01). "The Bakerian Lecture: Experiments and calculations relative to physical optics". Philosophical Transactions of the Royal Society of London. Royal Society of London. 94: 1–16. doi:10.1098/rstl.1804.0001
- Arumugam, Nadia. "Food Explainer: Why Is Some Deli Meat Iridescent?". Slate. The Slate Group. Retrieved 9 September 2013
- Chiao, R. Y.; Garmire, E.; Townes, C. H. (1964). "SELF-TRAPPING OF OPTICAL BEAMS". Physical Review Letters. 13 (15): 479–482. Bibcode:1964PhRvL..13..479C. doi:10.1103/PhysRevLett.13.479

Permissions

All chapters in this book are published with permission under the Creative Commons Attribution Share Alike License or equivalent. Every chapter published in this book has been scrutinized by our experts. Their significance has been extensively debated. The topics covered herein carry significant information for a comprehensive understanding. They may even be implemented as practical applications or may be referred to as a beginning point for further studies.

We would like to thank the editorial team for lending their expertise to make the book truly unique. They have played a crucial role in the development of this book. Without their invaluable contributions this book wouldn't have been possible. They have made vital efforts to compile up to date information on the varied aspects of this subject to make this book a valuable addition to the collection of many professionals and students.

This book was conceptualized with the vision of imparting up-to-date and integrated information in this field. To ensure the same, a matchless editorial board was set up. Every individual on the board went through rigorous rounds of assessment to prove their worth. After which they invested a large part of their time researching and compiling the most relevant data for our readers.

The editorial board has been involved in producing this book since its inception. They have spent rigorous hours researching and exploring the diverse topics which have resulted in the successful publishing of this book. They have passed on their knowledge of decades through this book. To expedite this challenging task, the publisher supported the team at every step. A small team of assistant editors was also appointed to further simplify the editing procedure and attain best results for the readers.

Apart from the editorial board, the designing team has also invested a significant amount of their time in understanding the subject and creating the most relevant covers. They scrutinized every image to scout for the most suitable representation of the subject and create an appropriate cover for the book.

The publishing team has been an ardent support to the editorial, designing and production team. Their endless efforts to recruit the best for this project, has resulted in the accomplishment of this book. They are a veteran in the field of academics and their pool of knowledge is as vast as their experience in printing. Their expertise and guidance has proved useful at every step. Their uncompromising quality standards have made this book an exceptional effort. Their encouragement from time to time has been an inspiration for everyone.

The publisher and the editorial board hope that this book will prove to be a valuable piece of knowledge for students, practitioners and scholars across the globe.

Index

www.ingramcontent.com/pod-product-compliance
Lightning Source LLC
Chambersburg PA
CBHW082038190326
41458CB00010B/3404